SOWING

By Leonard Woolf

History and Politics

INTERNATIONAL GOVERNMENT

EMPIRE AND COMMERCE IN AFRICA

CO-OPERATION AND THE FUTURE OF INDUSTRY

SOCIALISM AND CO-OPERATION

FEAR AND POLITICS

IMPERIALISM AND CIVILIZATION

AFTER THE DELUGE VOL. I

AFTER THE DELUGE VOL. II

QUACK, QUACK!

PRINCIPIA POLITICA

BARBARIANS AT THE GATE

THE WAR FOR PEACE

Criticism

HUNTING THE HIGHBROW

ESSAYS ON LITERATURE, HISTORY AND POLITICS

Fiction

THE VILLAGE IN THE JUNGLE

STORIES OF THE EAST

THE WISE VIRGINS

Drama

THE HOTEL

Autobiography

SOWING: AN AUTOBIOGRAPHY OF THE YEARS 1880 TO 1904

GROWING: AN AUTOBIOGRAPHY OF THE YEARS 1904 TO 1911

BEGINNING AGAIN: AN AUTOBIOGRAPHY OF THE YEARS 1911 TO 1918

DOWNHILL ALL THE WAY: AN AUTOBIOGRAPHY OF THE YEARS 1919 TO 1939

THE JOURNEY NOT THE ARRIVAL MATTERS:
AN AUTOBIOGRAPHY OF THE YEARS 1939 TO 1969

A CALENDAR OF CONSOLATION: A COMFORTING
THOUGHT FOR EVERY DAY IN THE YEAR

The author when a schoolboy

SOWING

AN AUTOBIOGRAPHY OF THE YEARS 1880 TO 1904

Leonard Woolf

HBJ

A Harvest/HBJ Book

HARCOURT BRACE JOVANOVICH, PUBLISHERS

San Diego New York London

Library of Congress Cataloging in Publication Data
Woolf, Leonard Sidney, 1880–1969.
Sowing.
Continued by Growing, an autobiography of the years 1904 to 1911.
Includes index.
1. Woolf, Leonard Sidney, 1880–1969. I. Title.
JA94.W6A3 1975 320'.092'4 [B] 75-12870
ISBN 0-15-683945-8 (pbk.)

Printed in the United States of America

First Harvest edition 1975

B C D E F

D. D. D.
Amico meo mihi
R. I. P.

CONTENTS

ILLUSTRATIONS

Chapter One

CHILDHOOD

I VERY rarely think either of my past or my future, but the moment that one contemplates writing an autobiography—and I am sitting down with that intention today—one is forced to regard oneself as an entity carried along for a brief period in the stream of time, emerging suddenly at a particular moment from darkness and nothingness and shortly to disappear at a particular moment into nothingness and darkness. The moment at which officially I emerged from non-existence was the early morning of November 25th, 1880, though in fact I did not personally become aware of my existence until some two or three years later. In the interval between 1880 and today I have lived my life on the assumption that sooner or later I shall pass by annihilation into the same state of non-existence from which I suddenly emerged that winter morning in West Cromwell Road, Kensington, so many years ago. This passage from non-existence to non-existence seems to me a strange and, on the whole, an enjoyable experience. Since the age of sixteen, when for a short time, like all intelligent adolescents, I took the universe too seriously, I have rarely worried myself about its meaning or meaninglessness. But I resent the fact that, as seems to be practically certain, I shall be as non-existent after my death as I was before my birth. Nothing can be done about it and I cannot truthfully say that my future extinction causes me much fear or pain, but I should like to record my protest against it and against the universe which enacts it.

The adulation of the deity as creator of the universe in Jewish and Christian psalms and hymns, and indeed by most religions, seems to me ridiculous. No doubt in the course of millions of millions of years, he has contrived to create some good things. I agree that "my heart leaps up when I behold a rainbow in the sky", or "the golden daffodils, beside the lake, beneath the trees, fluttering and dancing in the breeze", or "the stars that shine and twinkle on the Milky Way". I admit that every now and again I am amazed and profoundly moved by the beauty and affection of my cat and my dog. But at what a cost of senseless pain and misery, of wasteful and prodigal cruelty, does he manage to produce a daffodil, a Siamese cat, a sheepdog, a housefly, or a sardine. I resent the wasteful stupidity of a system which tolerates the spawning herring or the seeding groundsel or the statistics of infantile mortality wherever God has not been civilized by man. And I resent the stupid wastefulness of a system which requires that human beings with great labour and pain should spend years in acquiring knowledge, experience, and skill, and then just when at last they might use all this in the service of mankind and for their own happiness, they lose their teeth and their hair and their wits, and are hurriedly bundled, together with all that they have learnt, into the grave and nothingness.

It is clear that, if there is a purpose in the universe and a creator, both are unintelligible to us. But that does not provide them with an excuse or a defence. However, as I said, nothing can be done about it, and having made my protest, I must now think about my past. When I do think of my past and of the genes and chromosomes of my ancestors, for they after all are a highly important part of my past, I am a little surprised to see where they have landed me. I write this looking out of a window upon a garden in Sussex. I feel that my roots are here and in the Greece of Herodotus, Thucydides, Aristophanes, and Pericles. I have always felt in my bones and brain and heart English, and more narrowly a Londoner, but with a nostalgic love of the city and civilization of ancient Athens. Yet my genes and chromosomes are neither Anglo-Saxon nor Ionian. When my Rodmell neighbours' forefathers were herding swine on the plains of eastern Europe and the Athenians were building the Acropolis, my Semitic ancestors, with the days of their national greatness, such as it was, already behind them, were in Persia or Palestine. And they were already prisoners of war, displaced persons, refugees, having begun that unending pilgrimage as the world's official fugitives and scapegoats which has brought one of their descendants to live, and probably die, Parish Clerk of Rodmell in the County of Sussex.

For my father's father was a Jew, born in London in the year 1808. His name was Benjamin Woolf and he died in 1870 at the age of sixty-two in Clifton Gardens, Maida Vale. On his death certificate his occupation is given as "gentleman", but he was in fact by occupation a tailor who had done extremely well in his trade.[1] The first record of him and his business is found in the London directory of 1835, in which he appears as Benjamin

[1] I owe nearly all the accurate facts about my grandparents to my nephew, Cecil Woolf, who did a good deal of research into our ancestry.

Woolf, Tailor, 87 Quadrant, Regent Street. Sixteen years later he opened a second shop in Piccadilly, and four years later a third shop in Old Bond Street. He is described sometimes as "tailor and outfitter", and sometimes as "tailor, outfitter, and portable furniture warehouse", and in his shop at 48 Piccadilly as "waterproofer". He had seven sons and three daughters, and it is curious that three of the children married into the de Jongh family (my mother's family) and two of them married sisters, Louisa and Sarah Davis of Glasgow. His will is a formidable document covering many pages and he leaves considerable property, but with the proviso that none of it is to pass to any of his children if they marry out of the Jewish faith.

There still exist in London some tailor shops with WOOLF BROTHERS over the window—there used to be one in Holborn until Hitler dropped his bombs on it—and I used to think that perhaps they were the remains of my grandfather's business[1] which might well have made me a very rich man. It didn't, because, although my grandfather lived in a large house in Bloomsbury—I was once told that it was in Tavistock Square, but it may have been in Bedford Square—and had his seven sons, none of his sons went into the business. Like so many Jews of his class, he had an inordinate admiration for education and he educated his sons out of their class. I never knew any of my paternal uncles, but I doubt whether any of them benefited very much from their education, unless one can count one of them a success, for he certainly was, by all

[1] This is probably mere fantasy, but it is just possible because 1879 is the last directory to contain an entry for Benjamin Woolf and 1879 the first to contain an entry for Woolf Brothers.

accounts, an extremely brilliant and amusing scoundrel.

Two vast oil paintings of my paternal grandparents hung, during my childhood, on the wall of our dining-room, together with what was always referred to as "the little Morland" and a large pastoral scene, with sheep and goats, ascribed to Tenier. They—my grandparents—died before I was born, but their portraits which loomed so large over so many meals have indelibly impressed upon my mind their features and characters. I remember him as a large, stern, blackhaired, and blackwhiskered, rabbinical Jew in a frock coat, his left hand pompously tucked into his waistcoat, while she, who was born Isabella Phillips in 1808 and died in 1878 at the age of seventy, was the exact opposite: pretty, round cheeked, mild, and forgiving. Yes, it was all, no doubt, as it should be—the male forbidding and the female forgiving. She probably had a good deal to forgive, certainly from her children, all of whom, with the exception of my father and one of my aunts, must have been pretty tough people. The look of stern rabbinical orthodoxy in my grandfather's face was, I think, no illusion, for traditionally his family was just like that. His mother, my great-grandmother, we were told, used to walk to synagogue with hard peas in her boots in the evening of every Day of Atonement until she was well over seventy, and she stood upright on the peas in her place in the synagogue for twenty-four hours without sitting down until sunset of the following day, fasting of course the whole time. That in the Woolf family about the year 1820 was considered to be the proper way to atone for your sins. I feel a faint sneaking agreement with my great-grandmother, or rather I would if I had ever had a sense of sin. I presume that my unconscious is the usual cesspool of sadistic and masochistic guilt revealed by psycho-analysis, but I have never been able to detect in myself, even in childhood, a conscious sense of sin. But if there has to be this abominable doctrine of guilt and atonement, then I would approve of my great-grandmother's habit of doing the thing thoroughly.

My mother's family had none of the toughness and sternness of the Woolfs. The de Jonghs—her maiden name was de Jongh—were all of them rather soft. My mother was born in Holland of Jewish parents in Amsterdam. Her father was a diamond merchant, and the whole family migrated to London when she was still a child. I do not know why they migrated,[1] continuing the unending pilgrimage which, as I said, began in Palestine and Persia about 2,500 years ago. They must have been pretty prosperous, for the pilgrimage landed them in Woburn Lodge, that very pleasant kind of country house which, until just before the 1939 war, survived at the top of Tavistock Square beneath the caryatids of St. Pancras Church. Thus my maternal and paternal grandparents, my father, my mother, and I myself all lived in or practically in Tavistock Square. When my mother first went, as a child, to live in Woburn Lodge, in the road just outside was a turnpike which was only opened to residents in the Square by a ducal retainer who sat in a kind of summerhouse next to it. That must have been, I suppose, in the 1860's. When I went to live at No. 52 Tavistock

[1] My sister thinks that my grandfather often had to visit London and Paris on business, and that he liked London so much better than Amsterdam that he decided to transfer himself, his wife, and his ten children to England.

Square in 1925, the turnpike and the Duke of Bedford's retainer had gone, but the Square had not changed much otherwise since my mother's childhood. When she was over eighty, I walked with her one day across the Square to Woburn Lodge which was then empty and she stood in the melancholy little garden looking into the deserted rooms, which, she said, were exactly the same as they were when she was a young girl. But we were even then standing at the beginning of the end. For a year or two later they pulled down the north side of the Square and built up a very high and, architecturally, absurd building; and then, later still, in 1940 came the bombs which destroyed nearly all the south side, including No. 52, which, though we had moved to Mecklenburgh Square, we still had on lease from the Bedford Estate.

My mother's parents had ten children, most of them not only soft, but rather feckless, and they drained away most of my grandfather's prosperity by the time that I knew him. I knew both my maternal grandparents, as they lived to be nearly ninety. There was, in fact, something antediluvian about the de Jonghs, or at least they seemed to belong to a different century from ours. It was characteristic of them that my mother's nurse used to describe vividly to her the Napoleonic wars and how the French soldiers marched into the Dutch village where she lived as a young woman and were quartered on her parents. My grandparents, when I knew them, lived in a small house in Addison Gardens, and once a week we as children used to walk with my mother or with nurses or later with a governess from Lexham Gardens to Addison Gardens to have tea with them.

It was the cleanest house and they were the cleanest

people I have ever seen anywhere. My grandmother was always sitting by the window of the front ground-floor room in a black ebony chair which had an immensely high straight back rising several feet above her head. She never, so far as I know, stopped knitting, the needles going faster than I have ever seen in the hands of any other knitter. A large black cap was on the whitest hair, and beneath it was the round, pink face of an incredibly old Dutch doll. When she came to see us, the black cap was brought in a special basket made for the purpose. She was the kindest and roundest of women, and, though she had never, I think, read a book or had suffered from an abstract idea or had experienced the grinding of the intellect—which for most people is as unpleasant as the dentist's drill—somehow or other she seemed to have learnt to defeat fate. She was born round about 1800 in Groningen, a provincial town in the north-east of Holland, and in 1890 she sat in the window behind the lace curtains in Addison Gardens, having borne ten children and welcomed thirty grandchildren and having moved, imperturbable, from Groningen to The Hague and Amsterdam and thence to Woburn Lodge and Addison Gardens, passing on her way the whole of the nineteenth century.

Here is a curious example of a family tradition which must have had absolutely no foundation in fact: my grandmother's maiden name was Van Coeverden, and the family tradition was that it was one of her ancestors who in the eighteenth century discovered and gave his name to Vancouver Island. In fact the island was named in 1794 after an Englishman, George Vancouver, a captain in the British navy who first went to sea at the age of thirteen in the *Resolution*, the ship commanded by the great Captain

Cook. It is, however, possible that she had a good deal of non-Jewish blood in her ancestry. Some of her children and grandchildren were fair-haired and facially very unlike the "typical" Jew. She had a very nice, small, eighteenth century, black-and-white portrait of one of her male ancestors and he looked completely non-Jewish; the curious thing is that one of my brothers, who had fair hair, was so exactly like him that at first sight the picture seemed to be a portrait of him in eighteenth-century fancy-dress.

I am not one of Rousseau's latter-day disciples who believe in the nobility of the noble savage and in the wisdom of peasants, children, and imbeciles. From the ignorant I expect and I get ignorance and from the stupid, stupidity. But there are people—usually in my experience dogs or old women—extremely simple and unintellectual who instinctively know how to deal with life and with persons, and who display an extraordinary and admirable resistance to the cruelties of man, the malevolence of Providence, and the miseries of existence. My grandmother, sitting so upright in her ebony chair behind the white lace curtains, unconquered by the nineteenth century and her ten children, was one of them; she defied fate even in Addison Gardens. She died at the age of eighty-eight.

My grandfather was a very different type of person. He was a tall, gentle, rather silent man with a long white beard. No one could have mistaken him for anything but a Jew. Although he wore coats and trousers, hats and umbrellas, just like those of all the other gentlemen in Addison Gardens, he looked to me as if he might have stepped straight out of one of those old pictures of caftaned, bearded Jews in a ghetto, straight-backed, dignified, sad, resigned, expecting and getting over two or three millenia nothing but misery from the malignancy of fate and the cruelty of man, and yet retaining somewhere in the small of their backs or the cockles of their hearts a fragment of spiritual steel, a particle of passive and unconquerable resistance. In the house my grandfather always wore a brightly-coloured smoking cap, and I never saw him without a book and a cigar. I daresay that he too cherished in the small of his back and the cockles of his heart that particle of steel which alone enabled him to walk so upright, and alone can account for his survival, but I must admit that I never saw any evidence of it. Life and ten children and the nineteenth century seemed to have been too much for him, and, instead of defying, he had just yielded in silent melancholy to his fate. And at eighty-one, he was knocked down and killed in Walham Green by a horse-drawn omnibus.

The first experience of the misery of disgracing myself, so far as I can remember, came to me in my grandparents' house in Addison Gardens. The chairs and sofa in the sitting-room were covered in shining black horsehair. To sit on them in knickerbockers and stockings was, for a child of four or five years old, torture, for the stiff black hairs pricked you unmercifully on the bottom and behind the knees whenever you moved. I must have been about that age when one warm day the irritation in the lower half of my body, as I wriggled about in agony on the sofa, was too much for me and I had a violent impulse to cry and to make water at the same time. Unfortunately I fought heroically against both impulses until it was too late. The waters burst from me in two places simultaneously, pouring down my cheeks and my legs, and the old servant had to take me into the kitchen where I sat ignominiously wrapped in a blanket while my knickers and stockings were washed and dried.

I suppose some people would say that in this story, in this "misery of disgracing oneself", I have disproved what I said above, namely that I cannot remember ever having had a sense of sin. Isn't that feeling of disgrace the sense of sin? I would myself say no, and the distinction is, I believe, of immense importance in the psychology of the child and of the man or woman who grows out of the child. My father was a believing, but not an orthodox, Jew. He was a liberal Jew and a member of what was called the Reformed Synagogue. Jews of his generation and outlook were not much concerned with sin. They and their children escaped the psychological impact of crucifixion and redemption, of heaven and hell. It is true that there was the yearly Day of Atonement, but oddly enough my parents, unlike my great-grandmother, never seemed to connect it in any way with their or our sins. At any rate, I don't think I ever heard the word "sin" mentioned in our house; we were never beaten and hardly ever punished. Some things were, of course, wrong and some so terribly wrong it was inconceivable that any small Woolf would do them. We were, like all children, "naughty", particularly if we quarrelled or fought or did not eat up what was put on our plates at breakfast or lunch. The standard of behaviour, what was expected of a "little gentleman" or a "little lady" by nurses and governesses in the 1880's in Lexham Gardens, was pretty high. But both my parents were cheerful and kindly and good natured and took an extraordinarily optimistic view of God and his ordering of the universe. In consequence we were, as I remember it, extremely good children, yet not subjected to a perpetual stream of "Don't do this" or "Don't do that", and so with little, if any, *sense* of sin: indeed, until I went to school about the age of ten I was scarcely aware intellectually of the existence and importance of sin. To retain this innocence at St. Paul's Preparatory School or later at the private school in Brighton to which I went for two years was, of course, impossible, and I soon had an encyclopaedic knowledge of wickedness in man, woman, and child, both from the schoolmaster's point of view and that of the dirty and dirty-minded little boy. But by that time, I think, I had become inoculated against any feeling of personal guilt, for, though I often did things which I knew were considered to be wrong, I cannot remember ever to have felt myself to be a sinner.

This looking back at oneself through middle age, youth, childhood, infancy is a curious and puzzling business. Some of the things which one seems to remember from far, far back in infancy are not, I think, really remembered; they are family tales told so often about one that eventually one has the illusion of remembering them. Such I believe to be the story of how as an infant I fell into a stream near Oban which I heard so often that eventually it became part of my memory. What genuine glimpses one does get of oneself in very early childhood seem to show that the main outlines of one's character are moulded in infancy and do not change between the ages of three and eighty-three. I am sure that my attitude to sin was the same when I lay in my pram as it is today when I sit tapping this out on the typewriter and, unless

I become senile, will be the same when I lie on my deathbed. And in other ways when I can genuinely remember something of myself far off and long ago, I can recognize that self as essentially myself with the same little core of character exactly the same as exists in me today. I think that the first things which I can genuinely remember are connected with an illness which I had when I was about three. It was a very severe attack of scarlet fever which also affected my kidneys and in those days scarlet fever was a dangerous disease. I can remember incidents connected with the illness and I think they are genuine memories; they are so vivid that I can visualize them and myself in them.

The first is of a man coming into the room and applying leeches to my back. I insisted upon seeing the leeches and was fascinated by them. Twenty-five years later, one day in Ceylon during the rainy season, I was pushing my way through thick, wet grass in the jungle. I was wearing shorts and suddenly looking down I saw that my two bare knees were black with leeches. And suddenly I was back, a small boy of three, lying in bed in the bedroom high up in the Lexham Gardens house with the kindly man rather reluctantly showing me the leeches. I doubt whether in the intervening twenty-five years I had ever recalled the man with the leeches, but there in a flash the scene and the man and the leeches and my feelings were as vivid to me as the leeches on my knees, the gun in my hand, and the enveloping silence of the jungle.

When I look into the depths of my own mind (or should one say soul?) one of the characteristics which seems to me deepest and most persistent is a kind of fatalistic and half-amused resignation. I never worry, because I am saved by the feeling that in the end nothing matters, and I can watch with amusement and detachment the cruel, often undeserved but expected, blows which fate rains upon me. In another incident of my scarlet fever, which I think I do genuinely remember myself (though it became a family story), I seem to see this streak in my character already formed in the three-year-old child. At one moment my illness took a turn for the worse and I was, so it was said, upon the point of death. They called in Sir William Jenner, the Queen's doctor and a descendant of the Jenner who invented inoculation. He was a kindly man and I was fascinated by the shape of his nose. He prescribed a draught of the most appalling taste. I drank it down, but on his second visit—presumably next day—I sat up in bed with a second dose in the glass in my hand unable to drink it despite all the urging of my mother and Sir William. At last I said to them—according to my mother, with considerable severity—"If you will *all* go out of the room, I will drink it." I do not really remember that, but I do vividly remember the sequel. I remember sitting up in bed alone and the resignation with which I drank the filthy stuff, and the doctor and my mother coming back into the room and praising me. Sir William sat down on my bed and said that I had been so good that I would be given what I wanted. What did I want? "A pigeon pie", I said, "with the legs sticking out." "You cannot", he explained and his explanation was not unexpected by me, "be given a pigeon pie with the legs sticking out just yet, but you will be given one as soon as you are quite well. But isn't there something—not to eat—which you would like now?" I remember looking carefully into his kindly old face and saying: "I should like to pull your nose." He said that I might, and gently, not disrespectfully, but as a kind of symbol or token, serious but also, I believe, deep down amused, I pulled Sir William Jenner's nose.

But I must return to my father, Sidney Woolf.[1] He was sent by his father to University College School and afterwards, I think, to Kings College, London, or to University College in Gower Street. He was extremely intelligent and had a quick, powerful mind, so that he did very well at school. One of his elder brothers had become a solicitor and for a time my father joined him, but he had always determined to become a barrister. After some years as a solicitor, he was called to the bar. He was a first-rate lawyer and almost immediately successful; at the age of forty, as a Q.C., he made, I believe, over £5,000 a year. He was born at No. 87 Quadrant, Regent Street in 1844, married my mother in 1875, begat on her ten children, and died in 1892 after a few weeks' illness, at the age of forty-seven. I presume that, like every male, I was in love with my mother and hated my father, but I can find no trace of either the love or the hate in my memories, or indeed in my unconscious when, as occasionally happens, the id intrudes upon my ego. I was eleven when my father died. I admired him greatly and certainly thought that I was fond of him, and I think that he was both fond and proud of me, because as a small boy I was intelligent, reserved, and had a violent temper, and so in fact resembled him. He was certainly intelligent, reserved, and quick-tempered, but also very nervous and highly strung, and, though normally very kind, more intolerant of fools

[1] His name on his birth certificate is Solomon Rees Sydney Woolf, but he was always known as Sidney Woolf.

and their folly than almost any other man whom I have known. Though not an orthodox Jew, his ethical code of conduct was terrific, but he was not, in my recollection of him, either passionately on the side of righteousness or violently against sin. He must have been, I think, one of those rare people whose code of personal conduct is terrific, but whose morality is instinctive, springing from a delicacy or nicety of taste or aesthetic sensibility. This would explain why he was able unexpectedly to do without a sense of sin or the desire to punish sinners. He once said that in his opinion a perfect and complete rule of conduct for a man's life had, once and for all, been laid down by the prophet Micah in the words: "What doth the Lord require of thee, but to do justly, and to love mercy, and to walk humbly with thy God?" The words were inscribed on his tombstone in the grim and grimy cemetery in the Balls Pond Road.

I can remember the first time that I felt close to my father in a grown up way. I was only six years old, and I know that that was the case because it was the summer of the year of Queen Victoria's Jubilee, and as my father had a very long important case—I think it must have been an arbitration case—during the vacation, we had to take a house near London for our summer holidays so that he could get up and down to town easily. The house was at Kenley. When the rest of the family went off to Kenley, he had to stay on two nights in London, and for some reason I was chosen to stay on with him in Lexham Gardens. I felt terribly proud and important, particularly walking up Lexham Gardens by myself to the mews at the far end near Cornwall Gardens to tell Dennis, the coachman, what time the brougham would be wanted. We drove to the Temple and then walked across Lincoln's Inn Fields to a large room where the case was tried. It too must have been an arbitration case for we all sat round a long table and I sat on a chair next to my father. Mr. Bigham, who afterwards became a judge and Lord Mersey, was counsel on the opposite side, and there were heated arguments between him and my father. I thought him to be very rude and a most unpleasant man, and I was amazed when we adjourned for lunch and Mr. Bigham patted me on the head and we went off and lunched with him at the Rainbow Tavern, and my father and he were the best of friends. Later on in the afternoon they were at each other again hammer and tongues. Mr. Roper, my father's clerk, who seemed to be always arriving with a large red silk bag containing briefs, was very solicitous for my comfort. He must have been almost if not quite an albino, with the palest yellow hair and weak blinking eyes, and when one day, having come down with the red silk bag to a house in the country during one summer holidays, he was given tea, he amazed us by taking jam with his cake. Years afterwards someone might say: "Do you remember Mr. Roper eating strawberry jam with plum cake?", and heads were shaken over the aberration. The case went on for many hours and then in the evening we were driven back to Lexham Gardens and I sat up for dinner with my father. And next day, he drove us down to Kenley with its downs and white dusty road in a phaeton through Croydon and Purley.

My father worked so hard and so continually that we saw less of him than we and, I think, he would have liked. It was always exciting to be with him for he was extremely quick and lively in mind and body—when he was made a

Q.C. one of the legal papers described him as having "an eager and a nipping air". His energy was once almost the death of me. It was on my seventh or eighth birthday that he gave me a tricycle—it was the days before the "safety" bicycle was invented, and he and my elder brother, Herbert, already each had their tricycle. On Sunday we all three set off together on our tricycles from Lexham Gardens, along the Hammersmith Road, across the bridge to Barnes Common, to Sheen and Richmond Park. In those days, after the Castlenau Road you were practically in the country. It seems to me that it was a pretty long ride for a child of seven or eight who had not often been on a tricycle before. But, as it was afterwards discovered, there was something wrong in the bottom bracket of my new tricycle and I had simply to pound along using a good deal of force to make the beastly thing go at all. I can still remember the agony of grinding along the Sheen Road on the way back, the pain of exhaustion made worse by the disappointment in the present which I had so eagerly looked forward to. I managed to conceal my condition from father and Herbert, but when we got back to Lexham Gardens, we were met at the door by my mother, anxious to know how we had enjoyed ourselves. She raised a cry of consternation when she saw me stagger into the hall. It became a tale in the family saga, and she always said that, when she first saw me on the doorstep, my face was absolutely white except for the nose which was a flaming red. In the hall I collapsed and was carried away and put to bed.

My father's intellectual intolerance seemed to be roughly proportionate to his ethical tolerance. If a fool was anywhere in his neighbourhood, he tended to forget any idea of mercy or of walking humbly with or without his God. A stupid or silly remark would drive him frantic, and he showed little mercy to any man, woman, or child who made one. His irascibility when confronted by obstinate stupidity, at one time regularly every Sunday at lunch, produced a remarkable and to some a terrifying scene. Personally I looked forward to these scenes with astonishment, alarm, and at the same time a certain enjoyment. The culprit and victim was a cousin, the orphaned son of one of my father's sisters. In the 1880's a Victorian lunch in a Victorian family like the Woolfs was a formidable, but not altogether unpleasant ritual. It was eminently bourgeois, patriarchal, and a weekly apotheosis of the family. The change from the matriarchy of weekdays to the patriarchy of Sundays was very impressive to a small boy, and to me it was sympathetic. My father practically never stopped working. Every morning immediately after breakfast he was driven in his brougham from Lexham Gardens to Kings Bench Walk, where he had his chambers, and every evening at six the brougham fetched him back just in time for dinner at Lexham Gardens. After dinner and on Saturdays, if he was at home, he worked at his briefs. In the week, therefore, his children saw very little of him. Sometimes I was allowed to go into his dressing-room before breakfast to see him shaved, and sometimes my mother took me in the brougham to fetch him from the Temple—in the summer on such occasions we often stopped at the end of the Mall (or was it Birdcage Walk?) to drink a glass of milk from the Marsham Street cow who grazed of right in the corner of St. James's Park. But, as I was saying, this meant that we saw little of my father during the week, and Sunday lunch was a ceremony of some importance, for the whole family, capable of sitting upright and of eating roast beef, sat round the table.

I suppose that in the house in Lexham Gardens towards the end of the 1880's six of the nine children sat round the table at Sunday lunch with my father and mother. There was always an immense sirloin of beef, carved with considerable ceremony by my father. Bennie—as my cousin was always called—lived alone in London, and he had a standing invitation to our Sunday lunch; he was nearly always there. When I was ten, he must have been about twenty-three or twenty-four. He was almost, to look at, the comic Jew of the caricature, and he was that curious, but not very uncommon, phenomenon, the silly Jew who seems deliberately to exaggerate and exploit his silliness. He was the Jew so accurately described by one of the Marx brothers: "He looks like a fool and talks like a fool, but don't let him deceive you—he is a fool." Sooner or later, usually towards the end of lunch, Bennie would contrive to say something of inconceivable imbecility. My father with an effort would restrain himself and ignore Bennie. But Bennie was a masochistic moth who could not keep away from the devastating flame. He would turn with imbecile innocence to my father and ask him whether he did not agree with the imbecility. My father's fingers would begin to beat a nervous tattoo upon the tablecloth and all the little Woolfs fell silent round the table, staring apprehensively at the insensate Bennie. "But, Uncle Sidney," he would say, "Uncle Sidney, it is true, isn't it, that red-haired people in France are not taxed?" "No, it is not true, Bennie, and no one in the world but you would believe it." "But, Uncle Sidney—" and then my father would throw up his hands and let loose upon Bennie's head the torrent of his exasperation.

My mother was a good-looking young woman and we all liked to see her let down her hair, for it reached well below her knees and was extraordinarily thick. She must have been a perfect wife for she adored my father and yet was sufficiently different from him to make life interesting always for both of them. She adored her children and made life very interesting for them when they were small. The best hours of the day were between tea-time and my father's return from the Temple, for we spent them with her in the library playing, when we were quite small, and being read to later on when we were seven or eight or more. She was extremely lively and always ready for a joke both with us and with father. For instance, once when we were all away for the summer holidays in a house in Penmaenmawr, the rain came down in that solid, interminable, relentless way which seems peculiar to the grey mountains of Wales and Scotland. The feeling that there is no reason why it should ever stop, the conviction that it never will stop, induces in the human mind, particularly the child's mind, a feeling of complete despair. My parents and four or five of the elder children sat hour after hour in a largish sitting-room reading and looking out of the window at the grey sky and grey rain streaming down from the grey sky. Late in the afternoon my mother decided that something must be done. She dressed herself up in a black dress, with a black hat and a thick black veil and rang the front door bell. The servant came in and told my father that there was a lady on the door-step who asked to see him on urgent business. With some hesitation he agreed to see her. My mother was shown in and started off brilliantly with a long and somewhat confused story. Father did not recognize her, she played the part so well, and he began cross-examining her in his usual quick, incisive way. Suddenly she got a laughing fit and could not say a word to answer his questions. This was too much for his impatient nature and we rocked with laughter when he burst out: "My good woman . . ." My mother laughed so much that she had to snatch off the veil and reveal herself.

My mother had for my father and for his memory after his death something of the attitude which Queen Victoria had for Albert, though she was much less exaggerated and completely without the Queen's craziness. I suppose that Victorian matriarchs in widowhood tended to conform to this pattern in which a long life was dominated by the apotheosis of a dead husband. My mother was in many ways an ordinary middle-class woman, but twenty-five per cent. of her was a very individual and curious character. She lived to the age of eighty-seven or eighty-eight, and, if she had not insisted upon doing everything for herself—which meant that after the age of eighty she was always falling down and breaking a leg or arm—she might well have lived many more years. Physically, like most of her family, she was tough, though psychologically—again like them—soft. Or rather what made her a curious character was the strange mixture in her psychology of toughness and softness. To hear her talk you would sometimes have concluded that she was living in a world of complete unreality. And so up to a point she did. She lived in a dream world which centred in herself and her nine children. It was the best of all possible worlds, a fairyland of nine perfect children worshipping a mother to whom they owed everything, loving one another, and revering the memory of their deceased father. Nothing that actually happened, no fact, however black, however inconsistent with the dream, made her doubt its reality and its rosiness. That anyone, particularly one of her own children, should doubt or throw doubt on it was the one thing in life which really distressed her. She loved all her nine surviving children, but she loved me less, I think, than any of the eight others, because she felt me to be unsympathetic to her view of the family, of the universe, and of the relation of the one to the other. By nature she was a good-tempered and happy person, and we did not often have family rows or scenes, but every now and then we did have a terrific row, a most distressing scene, and it was nearly always caused by one of her children disturbing my mother's dream.

I remember once at dinner my eldest brother and I, and probably one or two younger brothers, were arguing vehemently about an incident which, according to the papers, had just taken place in one of our wars, probably the Boer war: a gunner of the R.H.A., rushing his gun up to a vital position, looked down and saw his brother lying wounded on the ground. For some reason, to stop his horses or swerve would have been contrary to orders or to the Gunners' tradition and therefore disastrous. Like a good soldier—so the papers said—he shut his eyes and drove over his brother. Was he right? My brother Herbert said he was wrong. I said he was right and gave an interesting account of the thoughts which would pass through my head and the arguments, military and moral, which would determine my action, if, in some future war,

I found myself unfortunately driving a gun over the wounded body of Herbert. We had become so interested in the problem that we had completely forgotten the presence of my mother and my sister. Both were in tears and almost in hysterics, and from about 8.30 until near midnight we tried without success to restore the damaged fabric of my mother's dream and calm the fury of feminine distress. My brother did not make the business easy. He was one of those persons who, with the best of intentions, can never leave well alone, and every time that my mother seemed to have got rid of the appalling vision of one of her sons driving a gun over his brother and had begun to recover the rosiness of her fairyland, Herbert, thinking to make everything doubly sure and convince mother that we had all meant the same thing all the time, would begin again with: "But, mother, you must see that", etc., etc., and we were instantly back where we started at 8.30 with mother and sister in tears.

The curious thing about my mother was that, although she lived in this dream world of rosy sentimentality and unreality, she was at the same time an extremely practical, sensible, hard-headed woman. When my father died, she found herself left with nine children, the eldest of whom was sixteen and the youngest three years old. Though my father had been making a considerable income at the bar, they spent nearly all of it, and suddenly at his death she passed from being very well off into a condition of comparative poverty. She had a little capital, but her income was quite inadequate to educate her nine children in the way that a barrister's children were habitually educated in the 1880's and 1890's. She was also saddled with a long lease of a very large house requiring seven or eight servants and an expenditure which she now could not possibly afford. My father must have been oddly careless about money, for at his own expense he had built a large wing on to the back of the house in Lexham Gardens, though he was the lessee, not the owner, of it. This was a fatal thing to do, for the house thus became much the largest in the street, and so it was now extremely difficult to let, being much larger and more expensive than the kind of house which the kind of people who wanted to live in Lexham Gardens expected to find there. It was indeed a white elephant to us, or an albatross hung round our necks, threatening to ruin my widowed mother and her nine innocent children.

For a year or two it depleted her capital, but just in the nick of time we got rid of it at some cost. She then showed great courage and sound sense. She took a much smaller house in Putney, into which she packed her nine children, a cook, a parlourmaid, and a housemaid, and she determined to spend the whole of her capital on educating her children, in the hope that by the time the money was exhausted they would be in positions in which they could maintain her and themselves. The gamble came off, but it would not have done so unless four of us had got scholarships at St. Paul's school and three of us scholarships at Cambridge. From my twelfth to my twenty-fourth year the menace of money hung over us all always and we had to be extremely careful of every penny; but my mother, though she had occasional panics, behaved on the whole with great common-sense, and though we all knew the risks we were running, we did not worry much about it.

The complete break in my life at the age of eleven, caused by my father's death, and the change from considerable affluence to the menace of extreme poverty had a curious effect upon me and in fact all of us. Looking back from 1960 to 1892, when my father died, I think that there was something to be said for the kind of life lived by the Victorian Woolf family in Lexham Gardens and by the many other similar bourgeois families in Bayswater and Kensington. It is, of course, condemned by Karl Marx and his all-red disciples, and it is because I condemn its economic basis and its economic effect upon other classes that I have been a socialist for most of my life. But the social standards of value in Lexham Gardens were very high, much higher than in any proletarian society today or in the proletarian section of a mixed class society. There is much which can be and has been legitimately said against family life on the grand scale, as developed by the middle classes of the nineteenth century: its snugness and smugness, snobbery, its complacent exploitation of economic, sexual, and racial classes. It had an innate tendency to produce the spiritual suburbanism which was the warp upon which so many superior novelists wove their stories between 1890 and 1914, a suburbanism which was a modern version of that lamentable philistinism which in previous generations had roused the sorrowful protests of Matthew Arnold and the fury and frenzy of Algernon Charles Swinburne. Yet it also had high psychological and aesthetic values, precisely those values which one feels so strongly in the family life as described in Tolstoy's novels. The actual relations between the human beings living in these large households and between the several households related by blood or friendship were, on the whole, in my remembrance extraordinarily human and humane. How much simpler everything would be if everything was either black or white, good or bad.

To return to the economic catastrophe of my father's death, it made us all a little more serious and mature than children between the ages of ten and fourteen are by nature. But it was an economic and materialist seriousness and maturity. I know all that there is to know about security and insecurity, of which much younger generations than mine have sung so many and such pathetic songs. I learned my lesson in 1892 before I was twelve years old. Before my father died, I—and, I think, the whole family—had a profound and, of course, completely unconscious sense of economic security, and, therefore, personally of social security. Money was not talked about or thought about or worried about; it was just there to be spent, not recklessly or extravagantly, but on things which ladies and gentlemen needed or wanted. And the social background, the house and servants and brougham and Sunday sirloin, which were based upon this invisible and unmentioned money, were accepted without question as stable and permanent, like the money. I was aware, at the age of eleven, that all this would send me to a public school, a university, and chambers in the Temple. This sense of economic and social security was, as I have said, innate and unconscious. It was followed suddenly in twenty-four hours by an acute and highly conscious sense of complete economic insecurity. Considering the tremendous reversal of fortune—which has always been assumed in literature to be the essence of tragedy—looking back, I am rather surprised that we did not take the whole thing more tragically. We did not worry much about the thing, but we became almost in a night economically serious and mature. In this we showed a good deal of sense.

We showed good sense because my experience convinces me that money is not nearly as important as we are inclined to believe. Until I reached the age of eleven we were very well off. For the next eleven years of my life we were extremely poor. Then for seven years I was comfortably off. When at the age of thirty-two I resigned from the Ceylon Civil Service, I had no money and no job, and, having married, Virginia and I had to work hard and be monetarily careful. After some ten years I found myself once more, as I had been in childhood, very well off. The point is that in all these economic vicissitudes, though money or its absence made a considerable superficial difference to one's way of life and the volume and quality of one's possessions, I cannot see that it ever had any great or fundamental effect upon my happiness or unhappiness.

In my view happiness and unhappiness are of immense importance, perhaps the most important things in life and, therefore, in an autobiography. It is curious that so little is known about them, particularly the happiness and unhappiness of children. I have pointed out in a serious book on politics, *Principia Politica*, that the apparently innate and profound unhappiness of the human infant, who will go into loud paroxysms of misery without provocation, is unknown in the young of other animals. This primeval pessimism of man must have great psychological and social importance, but autobiographically it is irrelevant, for, as far as I am concerned, I cannot remember anything about my infancy. At the time when my memories begin we were a cheerful and happy family of children, certainly above the average intellectually. But I can vividly recall two occasions when, at a very early age, I was suddenly stricken with an acute pang of cosmic rather than personal unhappiness.

My first experience of Weltschmerz, if that is what it was, must have come to me at the very early age of five or six. Behind the house in Lexham Gardens was a long parallelogram enclosed by the house on the north and on the other three sides by three grimy six-foot walls. It was a typical London garden of that era, consisting of a worn parallelogram of grass surrounded by narrow gravel paths and then narrow beds of sooty, sour London soil against the walls. Each child was given a few feet of bed for his own personal "garden" and there we sowed seeds or grew pansies bought off barrows in the Earls Court Road. I was very fond of this garden and of my "garden" and it was here that I first experienced a wave of that profound, cosmic melancholia which is hidden in every human heart and can be heard at its best—or should one say worst?—in the infant crying in the night and with no language but a cry. It happened in this way.

Every year in the last week of July or the first of August, the whole Woolf family went away for a summer holiday to the country. It was a large-scale exodus. First my mother went off and looked at houses. Then we were told that a house had been "taken". When the day came, six, seven, eight, and eventually nine children, servants, dogs, cats, canaries, and at one time two white rats in a bird-cage, mountains of luggage were transported in an omnibus to the station and then in a reserved "saloon" railway carriage to our destination. I can remember country houses in Wimbledon, Kenley, Tenby,

Penmaenmawr, Speldhurst, and Whitby which carry me back in memory to my fifth year. And I can remember returning one late, chilly September afternoon to Lexham Gardens from our holiday and rushing out eagerly to see the back garden. There it lay in its grimy solitude. There was not a breath of air. There were no flowers; a few spindly lilac bushes drooped in the beds. The grimy ivy drooped on the grimy walls. And all over the walls from ivy leaf to ivy leaf were large or small spider-webs, dozens and dozens of them, quite motionless, and motionless in the centre of each web sat a large or a small, a fat or a lean spider. I stood by myself in the patch of scurfy grass and contemplated the spiders; I can still smell the smell of sour earth and ivy; and suddenly my whole mind and body seemed to be overwhelmed in melancholy. I did not cry, though there were, I think, tears in my eyes; I had experienced for the first time, without understanding it, that sense of cosmic unhappiness which comes upon us when those that look out of the windows be darkened, when the daughters of music are laid low, the doors are shut in the street, the sound of the grinding is low, the grasshopper is a burden, and desire fails.

There is another curious fact connected with my passion among the spiders in the garden. Forty years later, when I was trying to teach myself Russian, I read Aksakov and the memories of his childhood. His description of the garden and the raspberry canes recalled to me most vividly my spider-haunted London garden and the despair which came upon me that September afternoon. I felt that what I had experienced among the spiders and ivy he must have experienced half a century before among the raspberries in Russia.

The second occasion on which I felt the burden of a hostile universe weigh down my spirit must have been when I was about eight years old. We had arrived in Whitby for our summer holidays and found ourselves in a large, new red-brick house on a cliff overlooking the sea. After tea I wandered out by myself to explore the garden. The house and garden were quite new for the garden was almost bare. Along the side facing the sea ran a long low mound or rampart. I sat there in the sunshine looking down on the sparkling water. It smelt and felt so good after the long hours in the stuffy train. And then suddenly quite near me out of a hole in the bank came two large black and yellow newts. They did not notice me and stretched themselves out to bask in the sun. They entranced me and I forgot everything, including time, as I sat there with those strange, beautiful creatures surrounded by blue sky, sunshine, and sparkling sea. I do not know how long I had sat there when, all at once, I felt afraid. I looked up and saw that an enormous black thunder cloud had crept up and now covered more than half of the sky. It was just blotting out the sun, and, as it did so, the newts scuttled back into their hole. It was terrifying and, no doubt, I was terrified.[1] But I felt some-

[1] It has been pointed out to me that Thomas Traherne seems to have had a similar experience. He writes in *Centuries of Meditation* (Third Century, No. 23): "Another time in a lowering and sad evening, being alone in the field, when all things were dead and quiet, a certain want and horror fell upon me, beyond imagination. The unprofitableness and silence of the place dissatisfied me; its wideness terrified me; from the utmost ends of the earth fears surrounded me. How did I know but dangers might suddenly arise from the East and invade me from the unknown regions beyond the seas? I was a weak and little child, and had forgotten there was a man alive in the earth."

thing more powerful than fear, once more that sense of profound, passive, cosmic despair, the melancholy of a human being, eager for happiness and beauty, powerless in face of a hostile universe. As the great raindrops began to fall and the thunder to mutter and growl over the sea, I crept back into the house with a curious muddle of fear, contempt, scepticism, and fatalism in my childish mind.

The child's mind and, since the child is father of the man, the man's mind are supposed to be formed very largely by religion and education. I find it very difficult in retrospect to discover what effect either had upon my mind and character. Both my parents were respectably religious. They believed in God. My mother went to synagogue on Saturday mornings fairly often, my father on the major feasts and festivals. They had us taught Hebrew by a rabbi who looked more like the traditional Jesus Christ than anyone else I have ever seen. He was an incompetent teacher and taught us a smattering of Hebrew, just enough to enable us to repeat a few Hebrew prayers. I cannot remember ever having actively believed in God though I suppose I must at one time have accepted his existence in a passive way. I think myself, though probably very few will agree with me, that my experience with the spiders and the thunder cloud destroyed any belief in God and religion which I may have had before. I know that it was not long after my fourteenth birthday that I announced that I was an unbeliever and would not in future go to synagogue, and I am sure that I had been contemplating this step for some time before I took it. When I solemnly announced to my mother that I no longer believed in Jehovah she wept, but her tears were not very convincing, I think, either to me or to her. She was genuinely distressed, but not very acutely; that I should repudiate the deity and refuse to go to synagogue caused a family sensation, but only a mild one which lasted a very short time.

As regards God himself, it is interesting to observe that by 1894 his position had already become precarious. No one, not even the believers, believed that he would take any steps against me for becoming an atheist. He had become as aloof, intermittent, and tenuous as a comet, and just as ineffective to impinge upon matter or to punish a sinner. Indeed no intelligent person any longer in practical affairs even considered the possibility of God intervening to reward the virtuous or punish the sinner any more than to bring rain to a parched crop or immunity from cholera and smallpox. People did of course still talk as if he could or might do so, but they acted as if he couldn't. In less than a century his position had suffered a change almost exactly like that of the British monarchy. He had become a constitutional instead of an absolute God. He got any amount of reverence and worship from his ministers and people; but all his powers and prerogatives had fallen from his hands into those of priests, the Archbishop of Canterbury or the Pope, of clergymen, churches, and chapels. All that remained to the deity was in fact caput mortuum.

Looking at the event from my point of view, I cannot see that loss of religious faith had any effect at all upon me morally either for good or for bad. I never suffered any of those torments of doubt and pain, remorse and horror, which have been described by their sufferers in many classical nineteenth century cases of men or women losing their faith. This may perhaps mean that my faith was always rather feeble or that I was never allergic to divinity. When I look into my heart and mind, I find a complete vacuum in certain places which in most other people seem to be full to overflowing. I feel absolutely no desire or necessity to worship. Indeed, I have an instinctive dislike of all gods and Gods, kings, queens, and princes. Unlike many people, I find it impossible to turn an ordinary man or an ordinary young woman into a myth of majesty and beauty, despite (or in part because of) the vast engines of modern propaganda, which in the press, the radio, and television, are employed so successfully to induce the masses to accept the miracle. The cry: "I must have a God and a faith, or I should have no hope", leaves me cold. I can get no comfort from believing what I want to believe when I know that there is no possible reason for believing it to be true. In fact, however, the universe would for me be a more comfortless place if it owed its origin and laws to one of the Gods whom man has invented than if it was merely the inexplicable phenomenon that on the surface (which is all we see and know) it appears to be. If Jehovah or almost any of the other major deities is our creator and ruler, the lot of man is hopeless, for he is subject to a "person" who is not only irrational, but cruel, vindictive, and uncivilized. The only tolerable Gods were those of the Greeks because no sensible man had to take them seriously. Of serious, major Gods only two have been civilized: Gautama Buddha and Jesus Christ. Buddha, however, is such an abstraction that he cannot be reckoned as one of the personal deities, and is to be regarded rather as one of the world's great philosophers, inventors of those fairy tales which we call metaphysics or rules of conduct which we call ethics. Christ seems to me to have been a great, but rather unpractical, man, who preached a civilized code of conduct and civilized way of life. If Europe had accepted Christ's Christianity and put it into practice, toning down or even rejecting some few rules of conduct prescribed in the Sermon on the Mount which are too utopian or civilized for human nature as we know it today, our history would have been less horrible and degraded and the world a far happier and far better place. Unfortunately, upon the civilized teaching of Jesus was grafted the dreadful doctrines of sin and punishment and that superstition which for thousands of years has haunted the savage and terrified mind of man—the belief that by killing or crucifying a God, a man, or a goat we can use their blood to wash away our sins. I am glad that I was never taught, as a child, this horrible doctrine of Crucifixion and Redemption.

Another thing which leads people to religion is the practice of praying. Here again I have never felt the slightest desire to pray to a God or to anyone or anything else. The whole business seems to me one of the oddest freaks in human psychology. It is easy enough to understand that if you really believe in a personal deity, and also believe that by the prescribed adulation, adjuration, and supplication you can induce him to do something to your advantage which otherwise he would not do or had forgotten to do—such as to make it rain over your fields, county, or country, or to destroy your enemy, or to forgive your sins—it is eminently sensible of you to pray to him or to hire a shaman, the Rector of the parish, or the Archbishop of Canterbury, as expert intercessor or professional go-between, to do the praying for you. But do nine out of ten of the people who pray or hire the professional prayers—do the professionals themselves, the parish priests and the Bishops and Archbishops, really believe today that they can induce God to make it rain, to destroy Germans or Russians, or to cure a child or a king of cancer or tuberculosis? I doubt whether they believe anything of the kind, and I simply cannot understand how they can go on year after year saying prayers which they know cannot have the slightest efficacy.

But there is an even odder phenomenon than this in the psychology of prayer. Years ago when I was in my early forties, and therefore pretty uncompromizing, I was one day eating my bleak plate of roast mutton at the unemotional dinner table in the tenebrous dining-room of 41 Grosvenor Road. In other words I was dining with the Webbs and I was sitting between Beatrice Webb and her sister Mrs. Henry Hobhouse. Somehow or other the subject of prayer turned up, and I said with some emphasis that I had never really prayed in my life; I had, of course, said my prayers as a child, the prayers which we were taught, like all well brought up, middle-class children, whatever the religious fold or sty we happened to be born into, to say morning and evening, but I had never prayed with the feeling that I was really addressing or asking someone something. I added that I could not understand how any intelligent person of the twentieth century could get himself into the frame of mind in which praying to God meant something to him. The two sisters fell upon me vigorously, the one from one side and the other from the other. I had met Mrs. Hobhouse very rarely, but I knew Mrs. Webb quite well and had the greatest respect and liking for her. She was one of the most intelligent persons I have known, but with some large blind spots in her intellectual and aesthetic vision. She told me that she habitually prayed with the utmost intensity and profound spiritual effect.

I tried that day at dinner and later in other conversations with Beatrice Webb to discover exactly what she prayed about and what the profound psychological effect upon her was, but I never got anywhere near an understanding. I do not think that she had any belief in a personal God or that she believed in anything which I should regard as religion, though she may have followed Shaw in the characteristic compromize, in deference to the scientific age, or substituting the Life Force for old bearded Jehovah. But can one pray to a Life Force? At any rate I failed completely to get Beatrice to explain what she prayed about or what the effect was. The nearest that I got to an explanation was that when she prayed she got the same kind of "release" or relaxation of tension which some people get from confession to a priest or a psycho-analyst.

The explanation of such a psychological idiosyncrasy as this is probably never simple and all sorts of contradictory thoughts and feelings were probably required to combine to make Beatrice Webb pray. At first sight she presented to one a facade of perfect poise and certitude, and with Sidney opposite her to catch and return with such precision the ball of conversation, it seemed fantastic to believe that any doubts or hesitation could ever assail the Webbs in Grosvenor Road. Occasionally one might say something to them which they would not discuss on the ground that it was "not their subject" and that they knew nothing about it—it is curious to remember that when I first knew them, which was before the 1914 war, foreign affairs was "not their subject". But even these flashes of ignorance only added to the impression of their omniscient certitude. For though neither of them was in the least arrogant, one was left with the impression that if the subject had any real importance it would not be "not their subject".

Sidney's facade was no facade at all; he was all the way through exactly what he appeared to be on the surface. He had no doubts or hesitations (just as he had never had a headache or constipation); for he knew accurately what could be known about important subjects or, if he did not actually know it, he knew that he could obtain accurate knowledge about it with the aid of a secretary and a card index. When Beatrice talked about religion or prayer, I never remember him to have taken any part in the conversation, though he seemed to follow it with attentive amusement. I am sure that prayer and God meant even less to him than they did to me. But you could not see much of Beatrice without realizing that, beneath the metallic facade and the surface of polished certainty, there was a neurotic turmoil of doubt and discontent, suppressed or controlled, an ego tortured in the old-fashioned religious way almost universal among the good and wise in the nineteenth century. I do not think tortured is too strong a word, for, if you watched Beatrice Webb when she was not the hostess, not talking, but attending only to her own thoughts, you would occasionally see a look of intense spiritual worry or acute misery cross her face. This deep-seated maladjustment is confirmed by her autobiography which reveals her as a woman of strong emotions with a profound conflict within herself between what she calls "the ego which affirmed" and "the ego which denied". She had too the temperament, strongly suppressed, the passions and imagination, of an artist, though she would herself have denied this. Her defence against these psychological strains and stresses was a highly personal form of mysticism, and in the consolatory process prayer played an important, if to me incomprehensible, part.

My attitude to prayer and religion appeared to irritate Beatrice, though in general she was, I think, fond of both Virginia and me. (During the war and not long before her death, she told Bobo Mayor that she would like to see us again and would come up to London for the day if Bobo would get us to lunch with her. We went and lunched with her, Mrs. Drake, and Bobo. She was more mellow and affectionate than I had known her before; she asked me what I was writing and, when I told her, characteristically said that I must read their *English Local Government*, Vol. IV, *Statutory Authorities for Special Purposes*, and a few days later she sent me a copy of the book. That was the last time I ever saw her.) To return to her attitude towards my attitude to religion, one Sunday when we were staying for a week-end with them at Passfield Corner, the conversation at lunch got on the subject of the teaching of religion in schools. When I said that I did not think it desirable that religion should be taught at all in schools, she was vehemently against me and carried the conversation from the luncheon table to the library. It was the first time that I realized to the full the strength of her passions and mysticism. She seemed to get angry that I mildly maintained my opinion, and marched up and down the room arguing almost violently. Indeed, up and down she marched faster and faster, and as she whisked herself round at each turn faster and faster, talking all the time, suddenly at one of the whisks or turns something in her skirt gave way and it fell on the floor entangling her feet. She stopped, picked it up, and holding it against her waist, continued her march up and down, never for a moment interrupting her passionate argument in favour of the teaching of religion in schools. Sidney and Virginia sat silent all through the discussion.

I do not understand Beatrice Webb's attitude to religion. It was peculiar to herself. As I said, I do not think she was religious or had a belief in God. Mysticism and scepticism were so nicely balanced in her that her mysticism was of the most generalized and intellectualized kind. She was intellectually too honest and austere ever to swallow or accept the religion of a Church, whether Anglican or Catholic. Difficult as I find it to understand her psychology, it is simple compared with that of intelligent intellectuals who, having attained that profound scepticism which is the religion of all sensible men, suddenly contrive to swallow in one gargantuan or synthetic act of faith the innumerable and fantastic doctrines and dogmas of the Church of England or the Church of Rome. I can understand how that Jew of Tarsus, a city in Cilicia, a citizen of no mean city, as he was accustomed proudly to insist, long ago on his way to Damascus with intent to persecute Christians, being a stern orthodox Jew, taught according to the perfect manner of the ancient Jewish law, suddenly in broad daylight seeing a great light from heaven surrounding him and hearing a voice say: "Saul, Saul, why persecutest thou me? I am Jesus of Nazareth, whom thou persecutest",—I can understand how Saul of Tarsus suffered instantly on the road to Damascus a complete conversion. But the belief to which he was converted was as simple as the belief from which he was converted. The great light and the voice from heaven convinced him that Jesus of Nazareth had not been a fraud, that he had really come to fulfil what the prophets and Moses had said would come, that he was the first man to rise from the dead—there he was after death appearing on the road to Damascus—and that it was his messianic mission to shew light to Jew and Gentile, teaching them how to repent and turn to God.

The psychology of this kind of conversion, whether through Balaam's ass or the visions of a St. Teresa, seems to me completely comprehensible. You believe already in some form of thaumaturgical religion and suddenly a new thaumaturgist or an apparent miracle converts you violently to some other form. But there is no comparison with this in the psychological somersault by which an intelligent sceptic acquires in one mouthful the encyclopaedia of amazing beliefs which successfully turn him into a Roman Catholic or a member of the Church of England. By what process of the mind or the emotions does one acquire sudden belief that the New Testament is a record of fact, that the Athanasian Creed is more certainly true and more significant than the multiplication table, and that those astonishing statements of the citizen of no mean city which the Rector of Rodmell reads to us when we assemble to bury one of our neighbours are not merely matters of fact, but also intelligible? There are one or two quondam sceptics whom I have known well, whom I still regard with admiration and affection, and whose somersault into a Church remains incomprehensible to me.

T. S. Eliot is the most remarkable. Tom, when we first knew him, was neither an Englishman nor an Anglican. I helped him to become an Englishman by becoming one of his statutory sponsors, and I am, I think legitimately, proud that I not only printed and published *The Waste Land* but had a hand in converting its author from an American to an English poet. I had no hand in converting him into an Anglican. In later years when he stayed with us at Rodmell, it filled us with silent amazement to see him go to early morning Communion at the village church. I could, if pushed to it, produce an intellectually adequate explanation of the psychological process which brought Tom into the respectable fold of the Church of England, but I have no sympathetic understanding of it as I have of many other mental states in which I do not actually share.

I have wandered forward from my point and my childhood, the point being the effect of religion and education upon me as a child. There is nothing more to be said about the effect of religion. As for education, what a strange, haphazard muddle it all seems to have been when one looks back upon it. The first teaching that I can remember to have received was at a girls' school in Trebovir Road, one of those many Kensington streets which were the waste land of Victorian middle-class dreariness. The school, at which my sister Bella had been for some years a pupil, was kept by a Mrs. Cole and it included a kindergarten presided over by a Mrs. Mole. Though the rest of the school was for girls only, co-education being in those days unknown, small boys were admitted to the kindergarten and entrusted to the incompetent Mrs. Mole. To the incompetent Mrs. Mole I was entrusted at the age of five or six. I cannot remember to have learnt anything at all in Trebovir Road except to take an early sexual interest in small girls. For besides the face of Mrs. Mole and the face of Mrs. Cole and the extremely low table at which we sat in the kindergarten, I can remember only two incidents. One was that I habitually sat illicitly holding under the table the hand of a small yellow-haired girl, and the other was that I somehow or other induced a rather older girl, with black hair, who was not in the kindergarten, to cause an open scandal by kissing me in the hall.

Whether it was considered that my education in other things was too slow or my education in sex too fast at Mrs. Cole's school, I do not know, but I was certainly removed from it pretty soon. My memory of what exactly happened to my education after that is hazy, at any rate for a time. We had for many years a governess living in the house, Miss Amy, who came from the Channel Islands and was bilingual in French and English. She taught us French and reading, writing, and arithmetic, all rather incompetently. Fräulein Berger came two or three times a week to teach us German, and I have a dim remembrance of other teachers, male and female, leaving an impression of being despised and dejected, insulted and injured, on my childish mind as they arrived weekly to teach us the piano, dancing, elocution, and other subjects of the same kind which everyone concerned, including the pupils, seemed to assume from the start to be hopeless and useless. It is extraordinary that people like my parents who attached great value to knowledge, books, and things of the mind should have been content, as they appear to have been, in the 1880's to provide such very poor teachers for their small children. Until I came into the hands of Mr. Floyd (of whom I will tell more in a moment) at the age of nine, I had never been taught anything to rouse my interest by anyone. Yet I am sure that my parents spent large sums, according to the standards of the time, upon our education. The people who taught us meant well and were all of them kind and decent to us, but they were themselves uneducated and quite uninterested in anything to do with the mind, and they therefore never interested me in anything they were teaching.

My nurse, who was with us for many years and brought us all up, had much less education than our governesses, but she was the first person to interest me in books and in the strange and fascinating workings of the human mind. She was a Somersetshire woman, born and bred on a farm, a rigid and puritanical Baptist. She read a Baptist paper every week from end to end and somehow or other she had got hold of a copy of de Quincey's *Confessions of an English Opium Eater*. This book entranced her; she read it again and again. I find it difficult to believe my memory when it distinctly tells me that Nurse Vicary used to give me a detailed account of what she read in the *Baptist Times* and often read aloud de Quincey to me, and that at the age of four or five I was quite an authority on the politics and polemics of the Baptist sect and often fell asleep rocked, not in a cradle, but on the voluptuous rhythm of de Quincey's interminable sentences whose baroque ornamentations must have been embellished by nurse's mispronunciations and her Somerset accent. But I had the deepest affection for her and for the opium eater, and she was the first person to teach me the pleasure of fear and thrill over public events, the horrors and iniquities of the great world of society and politics as recorded in the *Baptist Times* about the year 1885. I can still feel myself physically enfolded in the warmth and safety of the great nursery on the third floor of the house in Lexham Gardens, the fire blazing behind the tall guard, the kettle singing away, and nurse, with her straight black hair parted in the centre, and her smooth, oval peasant face, reading the *Baptist Times* or the visions of the opium eater. Just as the spider haunted garden remains in my mind as the primary pattern for all the waste lands and desolations into which I have wandered in later life, so the nursery with its great fire, when the curtains were pulled and the gas lit and nurse settled down to her reading, and occasionally far off could be heard the clop-clop of a horse in a hansom cab or four-wheeler, the nursery remains for me the Platonic idea laid up in heaven of security and peace and civilization. But though in the course of my life I have passed through several desolations of desolation more desolate than the garden with its grimy ivy and its spider webs, I never again found any safety and civilization to equal that of the gas-lit nursery.[1]

Outside the security of that nursery, terrible and terrifying things happened in the Kensington and London of fifty or sixty years ago. Hushed or whispered stories of Jack the Ripper, I think, penetrated into the nursery, and in my schoolroom days we were all terrified by a little woman, dressed all in black, who on foggy winter nights lurked in the Kensington streets, stabbed unsuspecting

[1] The nurse in the photograph of the Woolf family is not Nurse Vicary. She was a nurse who came and looked after infants in arms; I think her name was Mrs. Anselm.

gentlemen with a long knife, and then disappeared into the darkness and the fog. There is no doubt that in the eighties and nineties of last century under the prim and pious pattern of bourgeois life, just beneath the surface of society lay a vast reservoir of uncivilized squalor and brutality which no longer exists. It was a class reservoir, and the squalor and brutality welled up, in London at any rate, from those appalling slums inhabited by the "lower classes." It was when these dreadful drunken or savage creatures broke out for a moment from their lairs into the life of a small middle-class child that he first knew the paralyzing anguish of fear. I can still remember with the most sickening vividness some of the earliest occasions on which I learnt the agony and humiliation of unmitigated fear. The earliest of all is a memory of waking up in the middle of the night and hearing the shrieks of a woman pass along the Cromwell Road at the back of our house, pass along and fade away into the distance, leaving at last complete silence more terrifying even than the solitary shrieking. Next, standing on a chair at the dining-room window, watching the luggage being loaded on to the omnibus to take us all away on our holiday, and suddenly a drunken man in tatters, staggering about, trying to help with the luggage, cursing, swearing, becoming violent, and then finally the horrible sight of his vain struggle with a policeman and his being frog-marched away.

Thirdly, here is another scene. We are returning with nurses or governess down Earls Court Road having just passed the almost rural peace of Holland Walk and the sophisticated civilization of old Holland House. Suddenly out of a narrow side street, which led to one of the blackest of Kensington slums, two policemen appeared dragging a tall, raging and raving woman. They were followed by a small growling, but cringing crowd. Those who have never seen the inhabitants of a nineteenth-century London slum can have no idea of the state to which dirt, drink, and economics can reduce human beings. The men and women who surged or shuffled into the Earls Court Road behind the two policemen were, like the men and women whom La Bruyère saw in the fields in France, "animaux farouches". It is true that they had, like the seventeenth-century agricultural species, "une voix articulée" and, when they stood on their hind legs, human faces, so that, if nurse had read to me La Bruyère instead of de Quincey, I might have stood in the Earls Court Road of 1885, instead of in the France of 1685, and murmured "en effet ils sont des hommes". They were human beings, but they made me sick with terror and disgust in the pit of my small stomach, and the last scene, as the nurses hurried us away, is indelibly imprinted on my memory—the woman flung down in the middle of the road by one policeman, her battered black hat rolling away into the gutter, while the other drove back into their lairs the semi-circle of snarling "human beings". Such were the lessons in the sociology of classes which a child might learn in London streets about the time when Queen Victoria was celebrating the fiftieth year of her reign.

Looking back to that scene in a "respectable" Kensington street, I am struck by the immense change from social barbarism to social civilization which has taken place in London (indeed in Great Britain) during my lifetime. The woman, the policemen, the nurses, the small boy, the respectable passers-by averting their eyes—all these are inhabitants of a London and a society which has passed away. It can be counted, I suppose, as one of the miracles of economics and education. The slums and their unfortunate and terrifying products no longer exist. No one but an old Londoner who has been born and bred and has lived for fifty or sixty years in London can have any idea of the extent of the change. It is amazing to walk down Drury Lane or the small streets about Seven Dials today and recall their condition only fifty years ago. Even as late as 1900 it would not have been safe to walk in any of those streets after dark. The whole locality was an appalling slum, and its inhabitants, like all those of the innumerable slums scattered over London, were the "animaux farouches" described in the previous paragraph. They and their lairs, with the poverty, dirt, drunkenness, and brutality, have disappeared; the masses, which had terrified the bourgeoisie ever since they began to march from Paris to Versailles in October, 1789, have become the working classes and in England, at any rate, if a socialist dare say so, the working class has become almost indistinguishable, in its way of life, manners, and outlook, from the bourgeoisie. In the last forty years of my life I have got to know the life of the English countryside—in the south of England—as intimately as I know London—indeed, more intimately, for in London one knows intimately only a tiny fragment of its life—and I have seen the same process of profound social change, the emergence of a civilization out of a barbarism, take place in rural Sussex. In a later chapter I shall have something more to say of this.

I must return once more to my education. In Lexham

Gardens the children were divided between the nursery and the schoolroom. I do not remember at what age one was promoted to the schoolroom but I suppose it must have been round about the age of six or seven. Education began in the schoolroom which was presided over byMiss Amy, a tiny little Channel Islander. She looked exactly like a little robin, extraordinarily cheerful and sweet-tempered. I think we were all rather well behaved children, but sometimes I used to try her beyond endurance and she would burst into tears. She had a passion for jam puffs, which were sold at Andersen's bakershop in Earls Court Road, and, when Miss Amy was in tears, I always sneaked out to Andersen's and bought her a jam puff. The jam puff and "I'm sorry, Miss Amy" always brought immediate forgiveness. At some period of my childhood I was sent to St. Paul's Preparatory School. I cannot be quite certain when this took place, but I think it must have been in 1889, when I was nine years old. The only thing which I can remember about it is that I hated the place and was terrified by a boy who occasionally interrupted the relentless slowness of time and the narcotic boredom of the lesson by falling down in an epileptic fit. I cannot remember to have learnt anything at this preparatory school. I am astonished, when I recall the ten years of my education from 1889 to 1899, to find that the human brain could survive the desiccation, erosion, mouldiness, frustration applied to it for seven or eight hours every day and called education. I reckon that, before I went to Cambridge, I must have spent at least 10,000 hours of my short life sitting in some class-room, smelling of ink and boys, being taught by a gowned schoolmaster usually Latin, Greek, or mathematics, and occasionally

French or history. An immense number of those 10,000 hours was spent by me and, I think, the master in dense boredom. Of all my masters only two (or possibly four) were interested in what they were teaching and interested in making it interesting. My intelligence must have been considerable to have survived this process of desiccation and attrition.

I was only a term, or possibly two terms, at St. Paul's Preparatory School. In 1890, when I was ten years old, I was put under the care of a tutor who came daily to Lexham Gardens to teach me and my elder brother, Herbert. Mr. Floyd was a remarkable man, an eccentric who made the task of learning interesting. England, or rather Britain, breeds more eccentrics than most nations, and there is a national flavour to their eccentricity which is difficult to define or describe. In private life they are mostly bores, but they perform a useful purpose in leavening the heavy dough of English society or, to alter the metaphor, they help to keep the pores open to the flow of freedom. It was a good thing for a child of nine or ten to be taught by an eccentric like Mr. Floyd who had views of his own, unusual views, about most things and did what he thought right and proper unmoved by the misprision of his superiors or the ridicule of his inferiors.

Mr. Floyd was a tall, gaunt, long-legged man, very straight and upright, with thin greying hair and an absurd goatee. He had a large, wide-awake black hat which even in the street was more often in his hand than on his head. He had a curious look in his eyes of abstraction and ferocity. He instituted the following routine for us. After our breakfast Herbert and I walked to High Street Kensington station, where we met him. He then set off down

Wright's Lane to Lexham Gardens at a tremendous pace, Herbert on one side and I on the other. We had to run as fast as we could in order to keep up with him, his long legs striding out as in the pictures of the man in the seven-league boots, his head tilted up, the long thin hairs of his head and beard fluttering in the breeze, grasping in one hand an umbrella and in the other a black bag. Nearly everyone turned and looked at us with astonishment as we passed and small streetboys or cads, as we then called them (I don't think they exist in London today), hooted at us. Mr. Floyd paid no attention to anything like that.

As soon as we reached Lexham Gardens, we went straight through the hall to the small room overlooking the garden where we had our lessons. Mr. Floyd immediately sat down and we sat down one on each side of him. He put his large watch on the table, raised a ruler made of olive wood from Palestine, and said in a solemn voice, "Tacete", which is the Latin for "Be silent". Then for a quarter, a half, or a whole minute, as he chose, we had to sit absolutely motionless and silent. I still possess a little book in which I recorded from March 26, 1890 to May 26, 1891, the length of time each day I succeeded in being silent and motionless or failed. Mr. Floyd, in a beautiful hand, has headed the book "TACE" and has inscribed on the first page.

Qui non novit tacere, nescit loqui.
Stultus non novit silentium servare.

He has also written something in Hebrew, which is odd, because I am sure that he was not a Jew.

After "Tacete" we said the multiplication tables up to twelve as fast as we could. We did this daily until we succeeded in saying them without a mistake for three days running, each time within two minutes. This is not an easy thing to do and it took us quite a long time before we succeeded; it was considered that, having performed the feat, we knew the multiplication tables and need never say them again. There was a good deal to be said for Mr. Floyd's system, for, when I went to school, I found I was quicker than most boys in manipulating figures in simple arithmetic. He had also taught me to sit still and be silent on occasions, a rare accomplishment in a boy of ten. He taught me more than this. He had, I think, a genuine, if somewhat eccentric, passion for literature and he made one feel, even at that early age, that the books which we read with him—even Caesar *De Bello Gallico*—had something pleasurable in them, and were not merely instruments of educational torture. I have in my time been subjected, in the name of education, to so much mental torture, particularly the torture of the boredom of being taught by bored teachers, that I am grateful to Mr. Floyd for having made me dimly aware at the age of ten that lessons—things of the mind—could be exciting and even amusing. He had for some books the same kind of insatiable love as my nurse had for de Quincey. One of them was *Rasselas*, a copy of which he always carried in his pocket. We read *Rasselas* with him and he pointed out its beauties to us, but, unlike nurse's *Opium Eater*, it is a book which had and has no appeal to me. It seemed to me tedious and tiresome, but there must be something to it, because my mother, who sometimes came and listened to our lessons, became entranced by it. She bought a copy which was usually by her bedside and before she died she must have read the book dozens of times.

One of the pleasant things about Mr. Floyd was that he was one of those very rare people who never mind looking ridiculous. He taught us to play fives against a wall on the verandah and he also taught us singlestick, and as he was very tall and we were very small, the spectacle was extraordinarily absurd. But to see Mr. Floyd at his best was to see him reading Caesar with my canary sitting on the top of his head. I had a canary, called Chickabiddy, who was so tame that the door of his cage was never shut during the day and he used to fly about the room. I had taught him to come and perch on one's head if one called "Chickabiddy, Chickabiddy", and it used to be a game we played for two of us to stand at opposite ends of a room and make him fly backwards and forwards very rapidly from head to head. Mr. Floyd became very fond of him and he took a liking to Mr. Floyd's head. Mr. Floyd had a habit of walking up and down the room as he taught us, and Chickabiddy would sit the whole time on his head. The moment would come—eagerly awaited by us—when Chickabiddy would make a mess on the top of Mr. Floyd's head. If he was aware of the evacuation, he wiped the mess off with a piece of blotting-paper without interrupting the lesson. Sometimes he did not feel what had happened, and then I, as owner of the bird, said: "I am afraid, Sir, Chickabiddy has made a mess", and Mr. Floyd would say very politely: "Thank you, my boy", and wipe the mess off with the blotting-paper.

When my father died early in 1892, I was eleven years old and Mr. Floyd passed out of my life completely: he never came to see us again and I do not know what happened to him. He made a great impression upon me and I have a vivid memory of him, both physically and mentally.

I think he must have been a very humane and civilized man, but, young as I was, I felt that he was an unhappy and disappointed man. Well, he passed out of my life and I was sent to a boarding school at Brighton. It was Arlington House in Kemp Town, an expensive preparatory school of which the headmaster was a Mr. Burman. My brother Herbert had been sent there in 1891 when my father was still alive and we were well-off. A year later, when my father was dead, we were much too poor to afford the fees of Arlington House. But Mr. Burman, who was a stupid, but a very nice and generous man, insisted that Herbert and I should both come to the school at greatly reduced fees, and later on he did the same thing for my four younger brothers. He was one of the most ingrained conservatives I have ever known. Arlington House was a leading preparatory school in Brighton and full of sons of rich people. But Mr. Burman was so conservative that he would never change anything; in the first decade of the twentieth century things began to move and change even in middle-class education and Arlington House began to go downhill and eventually Mr. Burman had to give it up.

I was at Arlington House for two years from 1892 to 1894. The education which I received was no better and no worse than that usually given at the time to the sons of successful army officers, barristers, clergymen, and stockbrokers. I was taught to play cricket and soccer seriously by masters who thought both games of great importance. One of them, Mr. Woolley, was a first-class cricketer and a very good footballer, His attitude to cricket was that of an artist to his art. To be bowled or caught was pardonable; but to play an incorrect stroke or to cut or drive without "style" was, even though you might hit a four off the stroke, a crime. The whole school was lined up every day in the summer term and did "bat drill" for a quarter of an hour with Mr. Woolley, a handsome, dark, lean, graceful man, facing us with a bat in his hand, like a conductor before his orchestra. "Forward" or "off drive", he would say making the stroke perfectly himself, and the whole school would play forward or off drive, and he, like the great conductor, would spot even the smallest boy in the back row if he did not come perfectly straight forward or did not follow through in the drive in perfect style. I was quite good at cricket and in both elevens, and I learned from Mr. Woolley the seriousness of games, the importance of style, the duty when you go in to bat of making every stroke with the concentration which an artist puts into every stroke of his brush in painting a masterpiece. Since those days I have played nearly every kind of game from fives and bowls to golf and rugger, and I have played them each and all with the greatest pleasure. But Mr. Woolley's teaching had such an effect upon me that I cannot play any game unless I treat it seriously, i.e. each stroke or movement must be correct and above all you must aim at "style".

If we were taught to take games seriously by the masters at Arlington House, we were taught to take all other lessons not seriously. We were taught Greek, Latin, arithmetic, algebra, euclid, history, geography, French, and scripture. All the masters were hopelessly bored by all these subjects and so were we. Anyone seen to be good at lessons or rudimentarily intelligent was suspect both to masters and boys; to be a "swot", i.e. to take lessons at all seriously, was entirely despicable. I was then and have remained all my life a "swot"; I escaped the unpleasant consequences at Arlington House and later at St. Paul's, partly because I had a pretty violent temper and partly because I was sufficiently good at games to make intelligence and hard work pass as an eccentricity instead of being chastised as vice or personal nastiness. I must have been rather intelligent; otherwise I cannot see how I could possibly have learned enough from Mr. Burman and his assistants to win, as I did, a scholarship at St. Paul's in 1894.

The only thing I learned thoroughly at Arlington House, other than cricket, was the nature and problems of sex. These were explained to me, luridly and in minute detail, almost at once by a small boy who had probably the dirtiest mind in an extraordinarily dirty-minded school. I was at the time completely innocent and I had considerable difficulty in concealing from him the fact that it was only with the most heroic effort that I was preventing myself from being sick. However I soon recovered; one had indeed to develop a strong stomach in things sexual to stand up against the atmosphere of the school when I first went there. The facts are worth recording because they showed me for the first time at a very early age the enormous influence a few boys at the top of a school exercise upon the minds and behaviour of the masses below them. And what is true of a school is true also, I think, of almost every community or society of persons engaged in a common purpose or living in close relationship.

At the age of twelve I was not prudish, for I was much too innocent, and I do not think that I have ever been prudish after the nasty little X removed my innocence. But I have never known anything like the nastiness—

corruption is hardly too strong a word—of the minds and even to some extent bodies of the little boys in Arlington House when I first went there. I instinctively disliked it at the time and, when I look back on it, it rather horrifies me even today. It was entirely due to two or three boys at the top of the school. They set the unsavoury tone and dictated the unpleasant manners of all the rest of us. I think they were rather older than boys usually are in a preparatory school, being stranded there as they were too stupid to pass even the entrance examination for a public school. They therefore very soon left in a bunch and my brother Herbert became captain of the school and I succeeded him. Herbert was something of a Puritan and refused to allow what the previous "monitors" had encouraged. I followed his example, and, as we were both strict disciplinarians, when I left for St. Paul's in 1894 the atmosphere had changed from that of a sordid brothel to that more appropriate to fifty fairly happy small boys under the age of fourteen.

No attempt of any kind was made at this school to educate us to become intelligent and responsible members of English society. On the contrary, in so far as anything was done at all, it was calculated to make us unfit to live in any free, civilized society. We were taught nothing of contemporary events, and we were never given the slightest hint that what one learned could have any relation to the life one lived and would have to live. There was not a corner or crevice in Mr. Burman's mind which was not obstinately conservative. The only comment that I remember him to have made on public events was continual abuse of Mr. Gladstone, whom he regarded as the author of all evil. One day when the school was walking back from Brill's Baths along the front, Mr. and Mrs. Gladstone drove by in an open victoria. All the way people recognized him and many waved or took their hats off, and he bowed continually and took his hat off to them. He did not look at all the kind of criminal anarchist and traitor whose portrait Mr. Burman had drawn for us. My instinct has been, from a very early age, to disbelieve anything which I am told "on authority" or at the least not to believe it. At the age of thirteen, I think I had already seen far enough through the headmaster to accept everything he said, except on the subject of Latin verbs and the like, with some reserve, and the sight of Mr. Gladstone's eagle-like eminence sunning itself in the victoria confirmed my silent determination, since Mr. Burman was a conservative, to be myself a liberal.

One of the boys at the school was a grandson of John Bright. Mr. Burman never tired of gibing at him for having such an abominable grandfather. This was typical of what the masters at a first-class bourgeois preparatory school thought funny in the eighteen-nineties. But for pedagogic lack of humour the following is hard to beat. The French master at Arlington House was a M. Marot who claimed to be descended from a long line of Counts. He used to ride a horse which was peculiarly angular, and to our eyes he rode very badly. One day when returning from cricket we saw him riding on the front, and, meeting him later in the school, my brother outrageously said to him: "We saw you riding the old cow, Sir." M. Marot, who went purple in the face when angry, solemnly punished Herbert by making him write out 500 times: "I must not call M. Marot's horse a cow."

I wish I could recall vividly what it felt like to be a boy

of twelve or thirteen at a private school in Brighton in 1893. I have a dim remembrance of it and what I do remember is not at all like the usual picture presented to us by adults, whether parents, educationists, or novelists. There were intervals of terrific energy and high spirits, when, for instance, one was playing games or the whole school was romping about the garden or the gym. Otherwise one seemed to live in a condition of almost suspended animation, a kind of underwater existence, for my mental world had for me something of the dim, green twilight which the physical world must have presented, I thought, in Jules Verne's *Twenty Thousand Leagues Under the Sea*. It was a dream world; but it was the actual world of school that one seemed to be dreaming half awake, and always with the feeling that one was just about fully to wake up. I wanted to wake up and, at the same time, was half afraid of doing so. Now at the age of nearly eighty I am doubtful whether I ever have.

One of the reasons for my mental twilight was, I am sure, that I wanted to use my mind, but practically nothing was done to help one to do so—indeed, for the most part one was discouraged from doing so. At home, my mother encouraged us from an early age to read "good" books, Scott, Dickens, Thackeray, but it is a remarkable fact that until the age of sixteen, when at St. Paul's I got into A. M. Cook's form, none of my teachers, except Mr. Floyd, ever suggested to me that it was possible to read a work of literature or other serious book for pleasure.

I was not unhappy at my private school; indeed, I was usually quite happy, but it was the happiness of someone only half awake. Looking back, out of the welter of dimly remembered things I can recall a few which I enjoyed immensely. First and most important, food—to be taken out and given a lunch of steak and kidney pudding and ices at Mutton's on the front, after weeks of the rather nasty school food, was marvellous, and I can still recall the deliciousness of a large, hot Cowley bath bun which we were allowed to buy after bathing in Brill's Baths. Then sights—there was a clump of valerian in the garden and on a hot summer day one could watch the humming-bird hawk moths, two or three at a time, come to it. I remember too another entrancing sight connected with butterflies. In the spring we were always taken in coaches for the school treat to Laughton to have a picnic and wander about the woods. And down the glades, which in recollection seem to me to have been carpeted, as they now never are, with spring flowers, glided in the dappled sunlight, with that extraordinary velvety flight of theirs, dozens of the Pearlbordered Fritillary. I can also recall that it was at Arlington House that I first experienced intense pleasure connected with reading. In very bad weather and in the late afternoon of Sundays, the whole school sat in the big schoolroom in silence and read books which one could take out of a large cupboard, containing the "school library". Under the gas jets, on winter evenings, a great fire burning in the huge fireplace, in the silence and comfortable fug, I suddenly found myself transported from the rather boring and always uncertain life to which one had been arbitrarily and inexplicably committed, to the strangest, most beautiful, and entrancing world of *Twenty Thousand Leagues Under the Sea* or *The Log of the Flying Fish*. There is no doubt that I then experienced some of the exquisite pleasure, some purging of the passions, that later came to me, as to Aristotle, from more orthodox literary masterpieces.

In the autumn of 1893 I went to St. Paul's School in West Kensington, plunging with a shiver into a much larger and tougher world than I had known hitherto. There I at once began to develop the carapace, the facade, which, if our sanity is to survive, we must learn to present to the outside and usually hostile world as a protection to the naked, tender, shivering soul. At least, I suppose this is true. I have never known anyone who had no carapace or facade at all, but I have known people who had extraordinarily little, who seemed wonderfully direct, simple, spiritually unveiled. They may be highly intelligent and intellectual, but this nakedness of the soul gives them always a streak of the simpleton. They are, indeed, the simpletons—Koteliansky used to translate the Russian word as "sillies"—the "sillies" whom Tolstoy thought were the best people in the world. There was something of the "silly" in Virginia, as I always told her and she agreed, and there was a streak of the "silly" in Moore. Obviously there is something remarkably good in these streaks, and perhaps if anyone had the courage to be a complete "silly", to have no facade at all, he would get on just as well as, or better than, the tortoises, the timid souls who live their whole lives behind a shell or mask.

I am afraid that there was never a touch of the "silly" in me and I soon developed a carapace, which, as the years went past, grew ever thicker and more elaborate. The facade tends with most people, I suppose, as the years go by, to grow inward so that what began as a protection and screen of the naked soul becomes itself the soul. This is part of that gradual loss of individuality which happens to nearly everyone and the hardening of the arteries of the mind which is even more common and more deadly than of those of the body. At any rate, I certainly began to grow my shell at St. Paul's about the age of fourteen, and, being naturally of an introspective nature, I was always half-conscious of doing so. What the facade hides or is intended to hide in other people can rarely be known with certainty and the psycho-analysts would probably hold that it is even more difficult to know what lies behind one's own. I suspect that the male carapace is usually grown to conceal cowardice. Certainly in my own case, I believe, the character which I invented to face the world with originated, to a very large extent, in fear, in mental, moral, or physical cowardice. It was the fear of ridicule or disapproval if one revealed one's real thoughts or feelings, and sometimes the fear of revealing one's fears, that prompted one to invent that kind of second-hand version of oneself which might provide for one's original self the safety of a permanent alibi. When I said above that I was half-conscious of doing this, I did not mean that I did it deliberately; I did it instinctively, but, being introspective, was half-conscious that a mask was forming over my face, a shell over my soul.

I was five years at St. Paul's, from 1894 to 1899, when I went up to Trinity College, Cambridge. My education, in the technical and strict sense of the word, began at the beginning and ended at the end of those five years at school, for though I learned many and very important things in my five years at Cambridge, they were not the educational things which schools and universities are expected to teach one. The education which one received at St. Paul's in the last decade of the nineteenth century was Spartan in its intellectual toughness and severity. It was devised and enforced by the highmaster, F. W. Walker, a most curious and alarming man. He was a short but solid man, with a red face, rather bloodshot eyes, a straggly beard, a very wide mouth showing black teeth, blackened by the perpetual smoking of strong cigars. He had a deep and raucous voice of immense volume, and he usually roared with it as though in a violent rage, so that one often heard the bellow of the "old man", as of an enraged bull, echoing down the corridors or through the hall. The "old man" had developed and encouraged in himself the one-sidedness and eccentricity which are the occupational diseases of schoolmasters; he acted his part with such conscious vigour that he was almost a stage schoolmaster. I do not know anything of his private life, but I should guess that he only cared for two things: first an amalgam of St. Paul's School and Greek and Latin, and second an amalgam of good food, good drink, and good cigars.

I was as a boy, and am now, concerned with only the first amalgam, for it determined my education and the equipment of my mind. Despite, or because of, his barbarity and fanaticism the old man became a great headmaster, for, whether the school and the education are judged to be good or bad, they had a character of their own and were created by him. His vision of the school and education was narrow and fanatical. The object of a public school was, in his view, to give the boys the severest and most classical of classical educations. He seemed to be interested only in the clever boys and his object was to turn them into brilliant classical scholars. I think that he and his son, whom he had succeeded in turning into one of the most brilliant of Balliol scholars, the winner of every kind of academic prize and honour, had a genuine love for scholarship and even for classical literature, but their love had become fused and lost in a mixture of classics and St. Paul's School, of scholarship and scholarships. For the one test of whether St. Paul's School was doing what the "old man" wanted it to do had become in his view the number and quality of the scholarships which the pupils year by year won at Oxford and Cambridge colleges. It was generally felt that something would be wrong with the universe and with the school if in any year St. Paul's did not win more Oxford and Cambridge scholarships than any other public school and, at the least, one Balliol scholarship at Oxford and one Trinity scholarship at Cambridge. And in the eighteen-nineties there was usually nothing wrong with the universe or the school.

In order to turn small boys into scholars the highmaster had devised an extraordinarily intensive system of teaching Latin and Greek. If you came to the school without a scholarship, you were shuffled off at once into an appropriate form, on the classical side if you seemed to be fairly bright, but if not so bright, on the army, science, or history side. If you had won a scholarship, you were not at once drafted into an ordinary form; you were put into what was called "the Hall". Physically it was the school hall in which the whole school assembled daily for prayers and occasionally for functions like the annual prize-giving, called the Apposition. I, with all the other scholars of my year, was immediately put into "the Hall" under Mr. Pantin, a kindly but melancholy master. There we sat for a whole term day after day, for the whole school day, doing Greek and Latin composition. We did absolutely nothing else. At the end of it, the foundations of the ancient Greek and Latin languages—their grammar, syntax, vocabulary—had been ground into me as thoroughly as the multiplication tables by Mr. Floyd. They had become part of the permanent furniture of my mind. And this process of laying the foundations of scholarship was carried out by Mr. Pantin under the personal and terrifying supervision of the highmaster. At least once a day, and sometimes more often, the doors of the hall would be flung open and with an ominous swish of his gown the "old man" would sail in and flop down, with a growl and a grunt, on the form next to one of us. He then with great care corrected his victim's work in a curiously thin and palsied handwriting, with growls and grumblings and occasionally a roar of rage. As he did this almost daily for a whole term, he got to know personally at once each boy who had won a scholarship and could judge his intellectual capacity, i.e. whether X was a potential Balliol or Trinity scholar and whether Y would never be likely to get anything better than an exhibition at one of the smaller colleges.

The highmaster's character and methods were early revealed to me by the following incident. One afternoon a few weeks after I had been handed over to Mr. Pantin and his educational machine for producing classical pâté de foie gras, the highmaster swept into the hall and sat down on the form by my side. He turned and looked at me with a terrifying leer which revealed a satyr's mouth full of black and decaying teeth. He did not say anything until he had finished correcting my work. When he had put the pen down, he turned and gave me another leer.

"Boy", he roared at me, with a roar, not of rage, but of good humour, "boy, your mother has been to see me. Your mother did not like me." He then patted me on the head and went off to another apprehensive small boy. When I got home in the evening I heard from my mother, almost in tears, the story of her interview with Mr. Walker. She was, not unnaturally, very proud of my having won a scholarship at St. Paul's and had asked for an interview with the highmaster in order to discuss with him my brilliant future. She had, I suppose, expected to receive congratulations on being the mother of such a clever small boy. Alas, her lamb was torn to pieces. The boy, said Mr. Walker, had been badly taught; his Latin was hopelessly bad and his Greek worse. He knew no grammar and no syntax; he could not do Latin or Greek composition, and his translation was not much better; it was doubtful whether anything could be done with him. After five or ten minutes of this kind of tirade, he paused, and my mother, only just restraining her tears, said: "But, Mr. Walker, what can I do?" "*Do*, Mrs. Woolf," roared the highmaster, "*do*? You've done enough." And he got up, walked to the door, and opened it for my mother to go out. The interview was over, and the highmaster had attained what was, no doubt, his object: my mother never again asked for an interview to discuss her son's future.

The potential scholar, having spent a term in "the Hall", was then drafted into the classical form which Mr. Pantin and the highmaster considered appropriate. Until you reached the classical VIII form, you also did for a few hours a week French and mathematics; when you got into the VIII, you did nothing but Latin and Greek. But if you were on the classical side, nothing was considered to be of the slightest importance but Latin and Greek. The classical fanaticism of St. Paul's in those days may be seen in this. I was, I think, put into the Upper IV for classics, when I escaped from "the Hall". Although I had got a classical scholarship, I was, in fact, better at mathematics than classics. I was, therefore, immediately moved up into the VII form for mathematics, i.e. the top mathematical form on the classical side. It was taught by Mr. Pendlebury, a first-class mathematician, who had written a first-class school text book. And there I sat under Mr. Pendlebury for, I think, three years, never learning anything more than I learnt in the first year, because there was no higher form into which I could be moved and I had to do what each yearly succession of boys did in mathematical Form VII on the classical side.

When I reached the classical VIII form, I did nothing but classics, as I have said. And the intensiveness of the St. Paul's system may be seen from this. When I got into the top form of all, the Upper VIII, three or four of the most promising boys (of whom I was one) were withdrawn from the ordinary work of the form for a whole term before they were due to take the Oxford or Cambridge scholarship examinations. Every day we went to the highmaster's house and sat with the highmaster's son, "Dick" Walker, the great Balliol scholar, who made us all day long translate aloud in turn, without preparation, Homer, Virgil, or some other classical work. In this way we read, so far as I can remember, straight through the whole of the Iliad and the Aeneid, some dialogues of Plato and some Tacitus. We did nothing else at all. It gave one, at any rate for a time, a considerable Latin and Greek vocabulary and some mastery of the art of translating at sight. "Dick" Walker had such a phenomenal memory that, when we were translating Homer and Virgil, he did not use a book; he knew it all by heart so accurately that he could correct us if we made a mistake without looking at the text.

I said, some pages back, that I wished that I could recall vividly what it felt like to be the small boy who left Arlington House, Brighton, for St. Paul's School, West Kensington, about the age of 14. I have no doubt that deep down within me, beneath the facade, the carapace secreted by my soul, and beneath the psychological sediment and sludge of sixty years, that little boy still exists intact, so vulnerable, sensitive, eager, nasty, and nice. But I cannot summon him, even like a spirit, from the depths into my consciousness; I can see or feel him merely as a very dim, rather melancholy, emanation of myself. The youth of eighteen who left St. Paul's School for Trinity College, Cambridge, in 1899 is the same and yet so different. It would be an exaggeration to say that I can recall vividly what it felt like to be he or that I can remember exactly how he developed out of the small boy in the five years between 1894 and 1899. But there is no need to try to call him up from the depths; he moves recognizably within me, in my heart, my brain, and (if I have a soul) in my soul. For in developing into what he developed into he developed into me.

I had walked into St. Paul's School in 1894 a small boy; I walked out of it in 1899 a young man. This passage from boyhood to manhood is in many respects the most difficult and painful period psychologically of one's life. The human caterpillar and chrysalis, infant and boy, emerges as butterfly or moth; in my own case, I may perhaps be said to have emerged as that appropriately named variety of moth, the Setaceous Hebrew Character. The metamorphosis is much more commonly painful—and more painful—than novels and autobiographies admit or depict. I can, of course, speak only for my generation, now old, dead or dying. The modern infant and child, because happier, may perhaps find the passage less difficult, but there are signs that even he does not find it easy. First, one experiences the iron, ruthless impact of society upon the eager, tender, naked ego, upon the "dear little fleeting soul", the animula, vagula, blandula as Hadrian called it. It would be difficult to exaggerate the instinctive nastiness of human beings which is to be observed in the infant and child no less than in middle or old age. To call it original sin is absurd, for it would mean that we accept as true metaphysical fairy tales or religious nightmares. It is safer to recall and state the bare facts without inventing explanations like the Platonic ideas, Allah, Jehovah, or Jesus Christ. The fact is that at the age of ten, I was a fully developed human being, mean, cowardly, untruthful, nasty, and cruel, just as I was at twenty, fifty, and seventy. And when I observed my companions' actions or caught a glimpse of their thoughts behind the masks of their faces or the curtain of their words, I recognized in them the same intimations of immorality. Yet at the same time there was in all of us—or nearly all of us—I am sure, that animula, vagula, blandula, the gentle, eager, inquisitive, generous, vulnerable guest and companion of our bodies which seemed to have little or no connection with that other tough guest and comrade of the same body. And it was this vulnerable inhabitant of our bodies over which the irresistible steam-roller of society pounded in whatever private or public school to which our parents happened to have sent us, flattening us all out in the image of manliness or gentlemanliness which our parents or lords and masters considered appropriate. Whether the Hyde is more real than the Jekyll, or vice versa, in most human beings, I do not know. I had a feeling, and still have it, that my animula, vagula, blandula, was somehow or other more real, more myself, than the nasty little tough who was, as I thought, deep down and usually out of sight. I daresay that both of these beliefs were illusions. To one of my brothers I was pure Mr. Hyde, though he never revealed this to me until he was over sixty and I over seventy in the bitterest letter which I have ever received.

Having read *Genesis* and its story of Cain and Abel, and later Freud and his elaboration of it, the terrible story of the murderous hatred (suppressed of course) of son for father, father for son, and brother for brother, I ought not to have been astonished by this letter. But I was, not so much because my brother so obviously hated me or had seen the so carefully concealed Hyde behind the Jekyll. What shocked me and saddened me was that I should have known someone intimately for over half a century and have liked him, and never in all those years been aware of his hatred and contempt of me. When in Ceylon I for the first time saw in the jungle what nature was really like in the crude relation of beast to beast, I was shocked and at first even disgusted at the cold savagery, the pitiless cruelty. But when I contemplate the jungle of human relations, I feel that here are savageries and hatreds —illuminated by Zeus, Jupiter, Jehovah, Christ, or Dr. Freud—which make the tiger and the viper seem gentle, charitable, tender-hearted.

Let me go back to St. Paul's School through which I passed on my road to puberty, manliness, and gentlemanliness. My brother's letter shows that my preceptors and guides failed to put my feet on the path which would have landed me in the inner circle of gentlemanliness. But in my journey from form to form and from birthday to birthday I passed inevitably to the other two destinations, puberty and manhood. Sexually the passage to puberty was almost always for my generation a painful and unpleasant business; it certainly was in my case. The first time I ever had violent physical sexual sensations was as a very small boy when, in bed with a cold, I was reading a book called, I think, *The Scottish Chieftains*. The sensations astonished me; they came upon me as I read the description of how one of the chieftains—can it have been Wallace?—dashed down a hill and flung himself—without impropriety, I am sure—upon a lady who was being carried in a litter. I was puzzled by the involuntary physical phenomenon; vaguely I thought it must be somehow or other connected with the cold in my head, but it is perhaps significant that, despite my innocence, I did not report the symptom either to my nurse or to my mother. The facts about copulation and the birth of children were explained to me, as I have said, by a small boy at my private school in the worst possible way and to some extent inaccurately—I was left in some doubt as to the sexual functions of the female navel—when I was twelve years old. I remained a virgin until the age of twenty-five; the manner in which I lost my virginity in Jaffna, the Tamil town in the north of Ceylon, I will relate in a later chapter. In the thirteen years of chastity and youth which intervened my mind and body were continually harried and harassed, persecuted and plagued, sometimes one might even say tormented and tortured, by the nagging of sexual curiosity and desire. How dense the barbaric darkness was in which the Victorian middle-class boy and youth was left to drift sexually is shown by the fact that no relation or teacher, indeed no adult, ever mentioned the subject of sex to me. No information or advice on this devastating fever in one's blood and brain was ever given to me. Love and lust, like the functions of the bowels and bladder, were subjects which could not be discussed or even mentioned. The effect of this was, I believe, wholly bad, leading to an unhealthy obsession and a buttoning up of mind and emotion.

This withdrawal of the self into the inner recesses of one's being behind the facade and the series of psychological curtains which one interposed between oneself and the outside world of "other people" seems to me, looking back, to have been one of the dominant features in the progress from childhood to manhood. I was not an unhappy youth and we were not an unhappy family. I have already told of the reversal of economic fortune which fell upon us owing to my father's death when I was eleven. When we got rid of the white elephant of a house in Lexham Gardens my mother took her six sons, three daughters, a cook, a parlourmaid, and a housemaid to a house in Colinette Road, Putney. It was an ugly Victorian house, but "detached", with a small piece of garden in front and a largish square garden with fruit trees behind. To get thirteen human beings into it was a squeeze and it seemed at first very small after the spaciousness of Lexham Gardens. Considering the squeeze and the reversal of fortune, we were an unusually amicable family and quarrels, though sometimes violent, were rare. I was third in the family and I think that the change from wealth to comparative poverty made the eldest three children prematurely serious and grown up. My mother told us exactly where we stood; at the age of thirteen I knew that I must think carefully before I spent a sixpence or even a sixth of sixpence and that my future depended upon my brain and its capacity to win scholarships.

All this gave us—or at least me—as children a kind of grown up seriousness. A sudden reversal of fortune when one is a child impresses upon one, though one is not conscious of it at the time, a sense of the precariousness of life, the instability of one's environment. I know in fact the exact moment when that sense of instability came to me for the first time in my life. To a Londoner the rhythm of London traffic is part of the rhythm of his blood and of his life. I was born to the rhythm of horses' hooves in broughams, hansom cabs, and fourwheelers clattering down London streets, and body and blood have never completely synchronized their beat to the whir and roar and hoot of cars. One of my earliest recollections is of lying in bed high up in a front room of the house in Lexham Gardens, night after night, listening to the clop, clop, clop of a horse in a carriage or hansom cab break the silence of the night as it came down the street past our house. Clop, clop, clop—somehow or other that noise from outside gave one a sense of security, stability as one hugged oneself together under the bedclothes.

Reversal of fortune—to be on top of the world and next moment to be floundering in a bottomless pit, to feel the ground give way under one's feet, the bottom fall out of one's world—the Greek Sophocles recognized as the essence of tragedy. It remains its essence whether in the cosmic tragedy of Oedipus or the parochial tragedy of an eleven-year-old boy in nineteenth century Kensington. The moment came to me in bed listening to the horse's hooves fading away down Lexham Gardens. I remember it as the night before my father died and that somehow or other—perhaps from overhearing the hushed voices of servants—I was aware that he was dying and that his death meant not only the disaster of his death, the loss of him, but also the complete break up and destruction of life as I had known it. And in this curious vision of the future I saw that we were going to be "poor". I say that I remember the moment as coming on the night before my father died, but it is possible, indeed probable, that my memory is mistaken, for it is strange that a child of eleven should have been able and allowed to know so much before the catastrophe. But whether it was the night before or the night after death entered, I know the sense of security and stability had suddenly vanished; I could no longer, listening to the horse's hooves, hug myself in the haven of the bedclothes. The bottom had fallen out of the life and the house in Lexham Gardens.

From that moment a kind of unchildlike seriousness came into my life, a sense of responsibility and of the insecurity of material things like houses, food, money. It did not make me unhappy or, after the first shock, worry me. We were, as I have said, a cheerful and united family, lively, energetic, adventurous. I had and still have a passion for any kind of game, from chess or bowls to cricket and fives. I was quite a respectable bat and could play a respectable game at tennis or fives. I therefore enjoyed that important side of private and public school life which was concerned with games. At home we used to play cricket for hours in the back garden with a tennis ball and elaborate rules for scoring runs. My eldest brother, Herbert, and I developed very early a passion for bicycling. He must have been about twelve and I eleven when he acquired on his birthday his first bicycle. It was before the days of pneumatic tyres and we took the incredibly heavy and clumsy machine out into Lexham Gardens in order to acquire the art of riding. After a few minutes' practice he allowed me to try my hand, or rather legs. The seat was too high for me and I could only just reach the pedals, but he gave me a shove and I went off with great speed along the gutter, such speed indeed that I collided violently with a lamp post and the bicycle split in two, the handlebars and front wheel going in one direction, the back wheel, seat, and myself in another. Later we became experts and connoisseurs, saving up our money to buy cycles from a famous cycle shop in Holborn. I got exquisite pleasure from a cycle with handlebars like ram's horns and yellow rims to the wheels. Every day I bicycled to school from Putney to Hammersmith. In the holidays Herbert and I went on bicycle tours. We cycled all over England incredibly cheaply, for we could not afford to spend more than a few shillings a day. Our first expedition was to Oxford, Stratford, Evesham, and the West Country; our longest was to Edinburgh. When I was sixteen, we took our cycles by sea to the north of the Shetlands and cycled down to Lerwick. No one cycled there at that time and we were looked upon as bold adventurers. But one rainy night we were taken in and given beds in a small farmhouse. We were sitting round the fire after supper when there was a knock at the door and there, to everyone's amazement, was another cyclist. He was a young Aberdonian travelling in soap. As we sat talking after he had eaten his eggs and bacon, he saw my father's crest on a bookplate in a book I had been reading. The crest was a wolf's head with the motto THOROUGHLY under it. "I have a crest too," he said in his strong accent, "ay, and a coat of arms. The crest is a cat's head and the motto is SANS PURR; what d'you think of that, lad?"

At the age of fifteen or sixteen, therefore, we did what most boys do and on the surface as boyishly. Yet, beneath the surface, the reversal of fortune had had, I am sure, a darkening and permanent effect. In my own case I can only describe it as this sense of fundamental insecurity, and a fatalistic acceptance of instability and the impermanence of happiness. This fatalism has given me a philosophy of life, a sceptical faith which has stood me in good stead in the worst moments of life's horrors and miseries. For just as, though I believe passionately in the truth of some things, I believe passionately that you cannot be certain of the absolute truth of anything, so too, though I feel passionately that certain things matter profoundly, I feel profoundly in the depths of my being that in the last resort *nothing matters*. The belief in the importance of truth and the impossibility of absolute truth, the conviction that, though things rightly matter profoundly to you and me, nothing matters—this mental and emotional metaphysic or attitude towards the universe produces the sceptical tolerance which is an essential part of civilization and helps one to bear with some decency or even dignity the worst of Hamlet's slings and arrows of outrageous fortune.

This premature awareness of the seriousness of life

accelerated my passage from childhood to manhood and increased the withdrawal of the self into the innermost recesses. Looking back I can see now that there was another thing which strongly encouraged that withdrawal. Though I was not conscious of it for many years, indeed not until I was a young man, from the first moment of my existence, perhaps even before I left my mother's womb, I must have been "a born intellectual". The reading of books gave me immense pleasure, but so did "work" or lessons. Teachers in the days of my childhood and youth practically never explained to their pupils *what* they were teaching. For instance, mathematics, particularly algebra, gave me great satisfaction, though I was never told and never understood until years afterwards at Cambridge when I read Whitehead's little book, *Mathematics*, what on earth it was all about. This satisfaction which I got from mathematics is, I think, closely related to the aesthetic pleasure which came from poetry, pictures, and, most of all, in later years from music. But there were also in it the curious ecstasy which comes from *feeling* the mind work smoothly and imaginatively upon difficult and complicated problems, the excitement of the ruthless pursuit of truth which, perhaps, never entirely leaves one, but which is so intense when one is very young, and finally that astonishing and astonished happiness, described by Keats, which comes to one when some new constellation of thought, some new vision of a profound truth swims into one's ken.

All the characteristics which I have just described are the stigmata of the incorrigible, the born intellectual. England for considerably more than 100 years has been the most philistine of all European countries. This, I suspect, is largely due to the public schools, which during the period gradually established a dominating influence on public life and imposed upon the whole nation their prejudices, habits, morals, and standards of value. The public school was the nursery of British philistinism. To work, to use the mind, to be a "swot", as it was called in my school days, was to become an untouchable (except for the purposes of bullying) in the hierarchy of the public school caste system. Publicly to have confessed that one enjoyed any of these things would have been as impossible as for a respectable Victorian young lady publicly to confess unchastity and that she had enjoyed it. Overtly the only standard of human value against which the boy was measured was athleticism. Use of the mind, intellectual curiosity, mental originality, interest in "work", enjoyment of books or anything connected with the arts, all such things, if detected, were violently condemned and persecuted. The intellectual was, as he still is widely today, disliked and despised. This attitude was not confined to the boys; it was shared and encouraged by nearly all the masters.

This contempt of our teachers for what they were teaching and for the boy who wanted to be taught was on the face of it remarkable, but it was really natural and inevitable. In the kind of school to which I went nearly all the masters had been educated themselves in public schools; so too, probably, had their fathers before them. Instinctively and unconsciously and unquestioningly they accepted the standards of value and practised the precepts of public school tradition. They therefore naturally despised the intellect and the arts and anything connected with them, and so any small boy who showed any unusual intellectual ability or interest. To be a swot was just as despicable in the eyes of the masters as in those of the boys. The headmaster of my private school, Mr. Burman, as I have said, was the kindest and most generous of men; indeed, I owe a great debt of gratitude to him, for after the reversal of our fortunes he took me and all my brothers one after the other at greatly reduced fees. But he was a Philistine of the Philistines, a dyed in the wool Tory, a pure and perfect product of public school tradition. By the age of fourteen I had learnt from him and the other fifty small boys about me that one of the most despicable of things was to be too intelligent—and that you had to be pretty unintelligent if you wanted to be not too intelligent. Every master who taught me until I reached the age of sixteen or seventeen accepted and inculcated the same doctrine and ethic.

From my very early years I have had in me, I think, a streak of considerable obstinacy. I was lamentably intelligent, and, as I have said, I liked to feel the mind work, I was a born swot for I enjoyed my work. I was, of course, not fully or definitely conscious of this or of the hostility towards it in the world around me, for in childhood and youth, though one feels acutely what goes on in one's own head and in the heads of other people and their often painful interactions, owing to inexperience and diffidence one rarely fully understands or acknowledges what is going on. So I felt, but only dimly understood, the hostility of Mr. Burman and of boys and masters at Arlington House and St. Paul's to what I now see made me a horrid little intellectual. But, because at the back of my mind and in the pit of my stomach I had this little hard core of obstinacy, I never accepted the standards of value of Mr. Burman and of my environment. I did not rebel against them or openly challenge them, but I learned very early, I think, to go my own way behind the shutters of my mind and to be silent about much which went on there.

Being quite good at games and thoroughly enjoying them, I was able to carry this off and escape the penalties which awaited an intellectual in English schools in the last years of the nineteenth century. I was never bullied and, unlike many of my future friends, was never actively miserable at school. But my modus vivendi with masters and boys was attained only by the concealment or repression of a large area of my mental life which had the highest significance for me, and that was how the withdrawal of the ego into inner recesses, of which I wrote above, was encouraged and increased.

I was sixteen before I met anyone among my companions or teachers who showed any sympathy with the side of my life which I had sedulously concealed. When I went up into the VIII form—I think it was the Lower Middle VIII—the master was A. M. Cook, a brother of the editor of *The Daily News*. Cook was an extremely cultivated man; everything about him was quiet but strong including his passion for the arts and his sense of humour. He spotted my inclinations and capacities, I think, owing to my English essays (he made us take the writing of essays very seriously). At any rate, he quite soon asked me to walk round the playground with him during the morning break and for the remainder of the time that I was in his form we always spent this quarter of an hour together. I owe an enormous debt to A. M. Cook. He talked to me not as a master to a pupil or as an adult to a boy, but as an equal to an equal, on the assumption that we both accepted the same standards of intellectual and artistic value and obligations of truth. His taste was both strict and catholic. He encouraged me to read very widely and at the same time always to exercise my own judgment upon what I read. When I went up into a higher form, he gave me a copy of Bacon's *Essays*, beautifully bound in pale blue leather by Zaehnsdorf and inscribed in exquisite handwriting: "L. S. Woolf first in written work in L.M.8. St. Paul's School 1897: from AMC." The choice of Bacon's curious prose in the pale blue and gold of the Zaehnsdorf binding was characteristic of Cook.

In my last year at school I twice came into contact with people who did not despise the intellect and the arts. G. K. Chesterton had been at St. Paul's and was six years my senior. E. C. Bentley, the author of *Trent's Last Case* and the inventor of the clerihew, and R. F. Oldershaw were his contemporaries and they had founded a small debating society which met on Saturday afternoons in rotation in the houses of the members. Bentley and Oldershaw had gone up to Oxford, and, when I knew Bentley first, he was President of the Oxford Union. They continued the debating society after they had left school. It was kept quite small with only eight or nine members and they elected two or three boys still at school. How they came to elect me I cannot remember, but I know that I was both surprised and flattered. G. K.'s brother Cecil was my contemporary and a member; two other boys still at school were, I think, elected at the same time as I was.[1] One was called Myers and the other d'Avigdor, and it is

[1] I rather think that there was a third, namely S. P. Vivian, who eventually became a distinguished civil servant, Registrar-General, with a knighthood.

amusing, in view of the subsequent violent anti-semitism of the Chestertons, to note that three out of the four boys still at school whom they elected to this very exclusive society were Jews.

It was a queer society. The Chestertons were regular attendants and Bentley and Oldershaw came in the vacs. I never liked Cecil Chesterton, partly because his physical appearance was so unprepossessing, and partly because even then he had a streak of that kind of fanatical intolerance which seems to be fertilized, not by profound convictions, but by personal animosities. Gilbert was a very different kind of person. The monstrous obesity from which he suffered in later life had not yet attacked him, but like Cecil, though to a much smaller degree, he was physically unprepossessing. Whereas Cecil seemed to have a grudge against the universe, the world, and you in particular, G. K. gave one the immediate impression of goodwill, particular and general. In those days he had already begun to make his name as a journalist by writing for *The Daily News*. Our debating society was almost entirely political. It sometimes debated a particular political subject and sometimes functioned as a "mock parliament". G. K. practically never enlivened us with the paradoxical brilliance for which he was famous as a writer. My memory of him is standing very upright at the table, tearing sheets of paper into tiny pieces and dropping them on the table, while he spoke at immense length on some subject like taxation or bimetallism or the Irish question. His speeches were full of facts and good solid argument.

I do not know what eventually happened to this society the very name of which I cannot remember. (It may have been called The Junior Debater or something like that.)

After I left school and went up to Cambridge I dropped out of it, if indeed it continued to exist, and I lost touch with the members, including the Chestertons. Though our little debating society had been so exclusively political and ignored the arts, my enthusiasm for which had been encouraged by Cook, my contact with G.K. and the other members did bring a new breath of intellectual fresh air into my school life. The atmosphere of philistinism at a public school in the last decade of last century was pretty heavy, hostile, menacing to any boy who neither in his beliefs nor in his desires accepted the philistine's standards. It was not just a question of differing in beliefs and tastes. I got on quite well with the boys in my form or with whom I played cricket, football, and fives, but it would have been unsafe, practically impossible, to let them know what I really thought or felt about anything which seemed to me important. It was therefore a surprising relief to find oneself on Saturday afternoons with five or six people to whom one could say what one thought and who accepted the same intellectual standards of value whatever our disagreement might be about other things.

In my last year at school I came across two other people to whom I could talk freely. They were both with me in the top form, and, as they went up to Oxford with scholarships, I never saw them again after I left school. They belonged to the class of persons of whom unfortunately one has come across so many in one's life, the universal rebels who, though they do not know it, rebel against the universe or capitalism or Mr. Smith because they have a personal grudge against something or someone (quite often Mrs. Smith). After 1917 very many of these unhappy persons were able to sublimate their private grudges and hatreds, the torture of real or imaginary inferiorities, in the public or oecumenical grudges and hatreds of the Communist Party. But when I was a young man, Karl Marx and the Russian communists had not yet invented the international political lunatic asylum of twentieth century communism in which intelligent people can, in the name of humanity, satisfy animosities and salve their consciences. In those days the inferiority complex had few public outlets and became a kind of spiritual ingrowing toe-nail. My two friends, whom I will call A and Z, had this kind of ingrowing toe-nail. They were virulent intellectuals contra mundum. They despised and, I think, hated practically everything and everyone at St. Paul's, but they had a genuine intellectual curiosity and love of literature. When they found that I had the same, though they despised me for playing cricket and fives and for being friendly with all sorts and kinds of boys, they welcomed me as a conversationalist. I used to go into the classroom after cricket and about a quarter of an hour before afternoon school every day and meet them to stand at the window arguing interminably about everything under the sun. Here too I felt that I could say what I thought or felt. As I said, I never saw them again after I had left school. A became, I believe, a clergyman, and Z committed suicide while still at the university. The Church or suicide, it will be observed, were to us in the 1890's what the Communist Party became to a later generation.

Chapter Two

CAMBRIDGE

I WENT up to Trinity College, Cambridge, in October 1899. As a scientific exhibit, whether for individual or social psychology, my mind was in a curious state. They had turned me at St. Paul's into a pretty good classical scholar. In fact, I was good enough to make those in authority think that I might carry off the blue riband of Balliol, Oxford, or Trinity, Cambridge. I was therefore taken out of the Upper VIII for a term or two, and with two or three other boys subjected to the peculiar system of classical cramming or stuffing at the hands of Dick Walker, the highmaster's son, which I have described in the previous chapter. He suddenly advised me to "sit" for the scholarship examination at Trinity, Cambridge, in March, 1899. I had expected to go in for the usual examination later in the year, for the March examination at Trinity was mainly for people already at the college and, though open to outsiders, was very rarely attempted by a boy still at school.

It is almost impossible, I suppose, in old age to remember at all vividly even the miseries—let alone the splendours—of youth. No experience, except the first stages of falling in love, has such a mixture of acutely splendid and miserable torture as that of being "a new boy". The first day at school was to me—and I think to the vast majority of the male animals called boys—terrible and terrifying, but also exhilaratingly exciting. You suddenly found yourself in a new, strange jungle, full of unknown enemies, pitfalls, and dangers. It was the feeling of complete loneliness and isolation which made the fear and misery so acute, and the depth of feeling was intensified by the instinctive knowledge of the small boy that he must conceal the fear and misery. And mixed into the misery was the splendour of adventure, the excitement of entering into the new, the unknown jungle.

This terrifying experience of being a new boy is, of course, not confined to one's first days at school. It may happen to one all through life, though naturally it becomes rarer as one grows older, and it is the privilege—or perhaps infirmity—of old age that it is highly improbable that you will experience it after, say, sixty unless you have the misfortune to find yourself in prison or a modern concentration camp. I can remember at least seven such experiences in my own life: three times at the three schools to which I went; the grim days of the examination at Trinity; the first days when I went up to Trinity; the first days in Ceylon in the Civil Service; and the day when I was called up for military service in the 1914 war and entered Kingston Barracks for a medical examination.

The obstinate resistance to misery in the human animal is very remarkable. It may be absurd, but I know that it is true, to say that I was never more miserable than in the few days of March at the Trinity examination. I was the only examinee not in the college; I knew no one; no one spoke to me; I had no idea of where anything was or what I had to do except that I was to go to the Hall for dinner and the examination, and to sleep in a strange and uncomfortable room. Considering the state of nervous tension in which I was, I think it was a miracle that I did well enough to win an exhibition and subsizarship which enabled me to go up to Trinity in the autumn with £75 a year. In the following March at the college exam I converted my exhibition into a foundation scholarship of £100 a year. It is of some social interest to note that in the five years I was at Cambridge I managed to live on £120 a year—the scholarship provided £100 and my family the additional £20. I had to be extremely careful and economical, but never found any serious difficulty in living with friends like Lytton Strachey, Thoby Stephen and Clive Bell, who were well off and spent considerably more.

I entered the new, unknown jungle of Cambridge University and Trinity College as a resident in October, 1899. My adolescent mind was, as I say, in a strange condition. I had intense intellectual curiosity; I enjoyed intensely a large number of very different things: the smooth working of my own brain on difficult material; playing cricket or indeed almost any game; omnivorous reading and in particular the excitement of reading what seemed to one the works of great writers; bicycling and walking; work and the first attempts to write—I had won the Eldon Essay Prize at St. Paul's, £20 and a gold medal (sold by me many years later to Spink for, I think, £15) for an essay on monarchy, and I know that I got considerable pleasure (and pain) from writing it; people and talk, particularly the kind of people and talk which I wrote about at the end of the last chapter and which I had met so rarely in the jungle of school life that I had little hope of finding them in the new jungle of the university.

When for the first time as an undergraduate I walked through the Trinity Great Gate into Great Court on my way to the rooms at the top of a staircase in New Court which was to be my lair for two years, I trod cautiously, with circumspection, with no exuberant hopes of what I should find here—for that was what my experience of the human jungle had so far taught me. I had already developed, as I have said, a fairly effective and protective facade or carapace to conceal the uneasiness, lack of confidence, fear, which throughout my life I have been able to repress but never escape.

A symptom and part cause of this psychological flaw is the trembling of my hands which I have had from infancy; excitement or nervousness increase it, but it is never entirely absent. It is hereditary, for my father had it —I remember how, as a small child, I noticed that, when he sat in the library reading *The Times* after breakfast, the paper and his hands perpetually trembled a little. Two of his brothers were afflicted with it and more than one of my brothers. In Ceylon it proved to be a slight nuisance in a curious way. When one sat on the bench as Police Magistrate or District Judge, one had to make notes of the evidence and write down one's verdict and the reasons for it. Normally the tremor does not affect my writing and I would go on quite happily recording the evidence and the pleading of the lawyers if any were engaged in the case. But if I found an accused guilty, almost always a strange, disconcerting thing happened. When all the evidence had been given and the lawyers had had their say a silence fell on the hot court, as I began to write my analysis of the evidence and my reasons for my verdict. I wrote away without difficulty, but again and again when I got to the words: "... and for these reasons I find the accused guilty of ... and sentence him to ...", my hand began to tremble so violently that it was sometimes impossible for me to write legibly and I adjourned for five minutes in order to retire and calm myself sufficiently to complete the sentence (in two senses of the word).

I used often to wonder what was the explanation of this ironical situation in which the judge, the head of the district, the white "hamadoru"[1] found his hand refuse to convict and sentence to a week's imprisonment the wretched Sinhalese villager, though he knew that he was legally guilty of the offence and that the sentence was a lenient one. Was it some primeval, subterranean qualm and resistance due to the unconscious consciousness that the judge was no less guilty than the bewildered man in the dock? Or was it a still more subtle subconscious dislike of the majesty of the law as embodied in the judge? My reason has never allowed me to nourish any sentimental illusions or delusions about the law and those who break the laws. I think that I was, by Ceylon standards, a good Police Magistrate and District Judge, always feeling that it was my main duty to temper justice and severity by common-sense, the yardstick of his judicial function for the judge being, not his personal tastes and distastes and ethical beliefs, but the maintenance of law and the laws in the interests of order. In the court at Hambantota they would not, I think, have said that I was a "lenient" judge. But I have always felt that the occupational disease of judges is cruelty, sadistic self-righteousness, and the higher the judge the more criminal he tends to become. It is one more example of the absolute corruption of absolute power. One rarely sees in the faces of less exalted persons the sullen savagery of so many High Court

[1] This was the title which Sinhalese villagers always gave to the white civil servant; it was said to be something like "My Lord".

judges' faces. Their judgments, obiter dicta, and sentences too often show that the cruel arrogance of the face only reflects the pitiless malevolence of the soul.[1]

Such speculation is probably nonsensical and the explanation is probably simple, namely that my hand trembles because in the depths of my being I am physically and mentally afraid. (I used to tell Virginia that the difference between us was that she was mentally, morally, and physically a snob, while I was mentally, morally, and physically a coward—and she was inclined to agree.)

Another curious phenomenon is that the tremor in my hands has always tended to become extreme if I have to sign my name before other people, particularly on cheques or similar documents. This was a nuisance when I was head of a district in Ceylon (I was Assistant Government Agent of the Hambantota District in the Southern Province in my last three years in Ceylon), because there was a vast amount of signing of one's name in this kind of way which one had to do. When I came back to England in 1911, I went to a very intelligent and very nice "suggestion" doctor in Wimpole Street, Maurice Wright, and asked him whether he could cure me. He told me that I

[1] The faces of high dignitaries in the Church of Rome and the Church of England often exhibit the same kind of sullen malevolence. Perhaps it is difficult to reach high office in the Law or in the Church without becoming a hypocritical and angry old man. To watch judges at work in the Old Bailey or in the Court of Appeal, when criminal cases come up, will often show that what is said above is not exaggerated. When I used to be summoned to serve on a jury at the Old Bailey, and when I heard Lord Hewart trying criminal cases in the Appeal Court, I was often appalled and disgusted by the arrogant barbarism of the judge. Lord Chief Justices seem peculiarly prone to this kind of infection.

had a somewhat rare nervous disorder, called "familial tremor" because it ran in families; it was very difficult to cure either by suggestion or any other method. Some time ago, he said, a man in a business firm in Bombay had come to him because he had to sign a large number of cheques and as soon as he began to do so, his hand trembled so violently as to make it almost impossible for him to write his name. Suggestion had made him slightly better, but had not cured him. Wright held out little hope of a cure, but thought it just worth while to give it a try. After five or six sessions, he said I was not suggestible and it would be a waste of my time and money to go on with the treatment. He told me that he found a good deal of variation in suggestibility in different professions and occupations; in his experience policemen were the most suggestible of all men. For suggestion to work, he wanted a person to relax but not to become hypnotized; policemen were so suggestible that they almost invariably became completely hypnotized.

Another way in which my trembling hand proved to be a nuisance was when I went out to lunch or dinner; I would clatter my knife and fork on the plate or spill the wine on the tablecloth. Bernard Shaw, who had noticed this, once told me that he had been to F. M. Alexander who had cured him of some nervous affliction and he strongly advised me to let Alexander deal with my tremor. I went to Alexander and he treated me for some time. He was a remarkable man. He was a quack, but an honest, inspired quack. He had himself been suddenly afflicted with a nervous disorder and had cured himself by discovering that his loss of muscular control was due to the fact that he had got into the habit of holding his head and neck in the wrong position. From this he went on to maintain that all sorts and kinds of diseases and disorders were due to people getting into the habit of holding their head, neck, shoulders, and spine in this wrong position and his cure consisted in training the patient by exercises to abandon the wrong and acquire automatically the right posture. He said he could certainly cure me and for some time I went to him two or three times a week for treatment at considerable expense. I think that if I had had the patience to go on with the treatment and do the abominable exercises, I might have been cured or at any rate very nearly cured. But I simply cannot bring myself day after day to do physical exercises or remember to hold my head in a particular position, and gradually I gave the whole thing up. But Alexander himself was an extraordinarily interesting psychological exhibit. I feel sure that he had hit upon a very important truth regarding automatic muscular control and loss of control and that his methods could cure or relieve a number of nervous disorders. So far he was completely honest and a genuine "healer" of the primordial, traditional type. What was fascinating about him was that, though fundamentally honest, he was at the same time fundamentally a quack. The quackery was in his mind and came out in the inevitable patter and his claim to have discovered a panacea. However, as I said, I think his method might have cured me, but I had not the necessary patience to persist with the business and resigned myself to go on trembling slightly all my life. It is one of the consolations of age that it diminishes one's perturbations and fears, and so even one's tremblings.

But I must return to Cambridge in the autumn and winter of 1899. I felt terribly lonely in my first few days

The de Jonghs, the author's maternal grandparents, on their wedding day and on their golden wedding day

The Woolf family in 1886; the author is sitting in the front row on the extreme right

Sidney Woolf, Q.C., the author's father

Marie Woolf, the author's mother, in the dress in which she was "presented"

The Shakespeare Society, Trinity College, Cambridge. In the back row Thoby Stephen is second from the left, and on the far right Walter Lamb, later Secretary to the Royal Academy. In the front row on far left is Lytton Strachey and on far right the author; second from right is R. K. Gaye mentioned on page 40.

Thoby Stephen, about 1902

Vanessa Stephen, about 1902

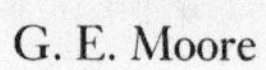

G. E. Moore

Duncan Grant and Maynard Keynes

Leslie Stephen and his daughter Virginia

Virginia Stephen, about 1902

at Trinity. I knew practically no one there or indeed at any other college. In my time at St. Paul's it was the fashion to enter for Oxford, not Cambridge, scholarships. None of my contemporaries in the Classical VIIIth came up with me and the only Pauline scholar of my year was Maxwell Garnett. As he was a mathematician, I scarcely knew him at school and barely knew him at Trinity. It was only twenty years later, when he was secretary of the League of Nations Union, that I got to know him well.

But suddenly everything changed and almost for the first time one felt that to be young was very heaven. The reason was simple. Suddenly I found to my astonishment that there were a number of people near and about me with whom I could enjoy the exciting and at the same time profound happiness of friendship. It began casually in what was called the screens, the passage through the Hall from Trinity Great Court to Neville's Court. I was looking at the notices on the board after dining in Hall and said something to a man standing next to me. We walked away together and he came back to my rooms. He was a scholar from Westminster, Saxon Sydney-Turner. Saxon was a very strange character with one of the strangest minds I have met with. He was immensely intelligent and subtle, but had little creativeness. In one of the university scholarship examinations they set us for Greek translation a piece from a rather obscure writer which had a riddle in it. Saxon won one of the scholarships and it was said that he was the only person to get the riddle bit right. It was characteristic of him. When, years later, crossword puzzles were invented and became the rage, he was a champion solver. And it was characteristic of him that he was a champion solver, never an inventor, of crossword puzzles and other mental gymnastics, including the art of writing. He had an immense knowledge of literature, but he read books rather in the spirit in which a man collects stamps. He would tell you casually that last night he had read for the second time in three weeks Meister Eckhart's *Buch der gottlicher Tröstung und von den edlen Menschen* much in the tone of voice in which a great stamp collector might casually remark—to *épater* his fellow collectors—that yesterday afternoon he had bought for 2*s.* 6*d.* in a shop in a back street of Soho two perfect specimens of a very rare 1*d.* Cape of Good Hope stamp. Later in life, when he was in the Treasury and lived in Great Ormond Street, he was an inveterate concert and opera goer in London and Bayreuth. He kept a record, both on paper and in his head, of all the operas he had ever been to. Normally with other people he was reserved, spoke little, and fell into long and unobtrusive silences. But sometimes he would begin to talk almost volubly about opera. He would tell you that last night he had been at Covent Garden and heard *Siegfried* for the thirty-fifth time. X had sung Brünnhilde; the great duet in the last act was quite good. X sang well and reminded him of Y whom he had heard sing the same part at Bayreuth, in 1908, Z being Siegfried, when he had been to *Siegfried* for the seventh time. The best performance he had ever heard of the opera was his twelfth, also at Bayreuth; Y was again Brünnhilde and there was the greatest of all Siegfrieds, W. The fourteenth time he saw the opera was . . . and so on.

The rooms which Saxon lived in for many years in Great Ormond Street consisted of one big sitting-room and a small bedroom. On each side of the sitting-room fire-place on the wall was an immense picture of a farmyard scene. It was the same picture on each side and for over thirty years Saxon lived with them ever before his eyes, while in his bedroom there were some very good pictures by Duncan Grant and other artists, but you could not possibly see them because there was no light and no space to hang them on the walls. As time went on, Saxon acquired more and more books and, since he suffered from a variety of ailments, more and more medicine bottles. His bookcases filled up and soon a second and third row, one behind the other, became necessary, and then piles and piles of books covered the floor. There were books upon the tables and chairs, and everywhere there were empty medicine bottles on the books, and the same two pigs, the same two sheep, and the same two dogs looked down upon, one presumes, the unseeing Saxon from the same two pictures on either side of the mantelpiece.

I was up at Trinity for five years. The first two years I had rooms in New Court; in the last three years Saxon and I had a double set of rooms in Great Court. It had one very large room on the first floor and two small bedrooms on the second. Saxon was a short, thin man with a very pale face and straw-coloured hair. He seemed to glide, rather than walk, and noiselessly, so that one moment you were alone in a room and next moment you found him sitting in a chair near you though you had not heard the door open or him come in. We saw very little of each other except in the evenings for he used to get up very late as a rule whereas I was up at eight. We hardly ever had a meal together for he ate very little and at the most erratic hours.

Both physically and mentally Saxon was ghost-like,

shadowy. He rarely committed himself to any positive opinion or even statement. His conversation—if it could rightly be called conversation—was extremely spasmodic, elusive, and allusive. You might be sitting reading a book and suddenly find him standing in front of you on one leg in front of the fire knocking out his pipe into the fire-place and he would say without looking up: "Her name was Emily", or perhaps: "He was right." After a considerable amount of cross-examination, you would find that the first remark applied to a conversation weeks ago in which he had tried unsuccessfully to remember the christian name of Miss Girouette in *Nightmare Abbey*, and the second remark applied to a dispute between Thoby Stephen and myself which I had completely forgotten because it had taken place in the previous term.

During the years we were at Trinity, Henry James was at the height of his powers, writing those strange, involved, elusive novels of his last period. We read *The Sacred Fount*, *The Wings of the Dove*, and *The Golden Bowl* as they came out. Lytton Strachey, Saxon, and I were fascinated by them—entranced and almost hypnotized. I don't know whether we thought that they were really great masterpieces. My enjoyment and admiration of them have always been and still are great, but always with a reservation. There is an element of ridiculousness, even of "phoneyness" in them which makes it impossible to rank them with the greatest or even the great novels. But the strange, Jamesian, convoluted beauty and subtlety of them act upon those who yield to them like drink or drugs; for a time we became addicts, habitual drunkards—never, perhaps, quite serious, but playing at seeing the world of Trinity and Cambridge as a Jamesian phantasmagoria, writing and talking as if we had just walked out of *The Sacred Fount* into Trinity Great Court. The curious thing was that, whereas Lytton and I were always consciously playing a game in talking or writing like Mrs. Brissenden and Mrs. Server, Saxon quite naturally talked, looked, acted, *was* a character in an unwritten novel by Henry James.

No human being can be quite as cynical, quite as ironical as facts. While I was in Ceylon—about 1908 or 1909, I suppose—Vanessa, Virginia, and Adrian Stephen went to Rye for the summer and Saxon stayed with them. Henry James was living in Rye then, in Lamb House, and there was also living in the town at the same time, Sydney Waterlow.[1] Sydney, who was a great friend of the novelist, told me that James was shocked by the "Stephen girls" or rather by their friends. James had known the Stephen children well from their childhood for he was an intimate friend of Sir Leslie Stephen and often came to the house in Hyde Park Gate when their mother was alive. When they came to Rye that summer, he had not seen them for a good many years. He was uneasy at not finding in them the standard of lady-like life and manners which belonged to Hyde Park Gate and the houses and their inhabitants in *The Wings of the Dove* or *The Golden Bowl*. But what upset him most was their friends, poor Saxon and Lytton Strachey, who also came to stay with them. Sydney repeated to me with gusto an interminable sentence in which by parenthesis within parenthesis and

[1] The late Sir Sydney Waterlow. Sydney's life was in some ways stranger than fiction. He was an infant prodigy at Eton and a brilliant classical scholar at Trinity College, Cambridge; later in the Diplomatic Service, his last post was Minister in Athens.

infinite reservations, involutions, and convolutions Henry James delicately, regretfully, hesitatingly conveyed his feeling that Saxon was small, insignificant, silent, and even rather grubby.

Nothing could have been more ironical than the situation there in Rye fifty years ago—the infinitely subtle author of *The Sacred Fount*, with his infinitely sensitive antennae, rendered completely insensitive and obtuse by the mist of social snobbery through which he saw life and people and out of which he often created his shadowy masterpieces. For in 1907 Henry James was in many ways a disappointed man. His reputation was high, but his readers were few. Like so many writers, and with a good deal more reason than most, he felt that the readers, the sales, the success which he knew he deserved evaded him. This saddened him and he was immensely pleased by the appreciation and admiration of younger people like Sydney Waterlow and Hugh Walpole. But Sydney and Hugh were extremely respectable young men, properly dressed, with the right hats on their heads, and carrying an umbrella at the appropriate moment. And now there was the novelist sitting in the same room with two of the most intelligent of the younger generation who understood and admired him far more profoundly, I think, than Hugh or Sydney did, and one of them, Saxon, was almost a creation of the novelist, a character in one of his novels. And all that the sensitive antennae recorded was that the young man was small, silent, and grubby.

All this, it should perhaps be added, did not permanently affect James's respect and affection for Leslie Stephen's family. After Virginia and I married in 1912, I acted for a short time as secretary of Roger Fry's Second

Post-Impressionist Exhibition at the Grafton Galleries. One afternoon Henry James came, and after Roger and I had shown him round the pictures—which he did not very much like—we took him down into the basement and gave him tea. When he realized that I had just married Virginia, he got up and shook hands with me a second time and made me a characteristic, ceremonious speech. It went on for quite a time and had many trailing and flowery sentences full of parentheses. But it showed, I thought, genuine kindliness and real feeling for Leslie Stephen and for the great beauty of his wife and daughters. I was amused to see that during tea, as he talked, he gradually tilted back his chair until it was balanced on the two back legs, he maintaining equilibrium by just holding on to the edge of the table. Now the Stephens had told me that when they were children and Henry James came to tea, or some other meal, which he often did, he had a habit of doing this when he talked. As the long sentences untwined themselves, the chair would tilt slowly backwards and all the children's eyes were fixed on it, fearing and hoping that at last it would overbalance backwards and deposit Henry James on the floor. Time after time he would just recover himself, but then indeed at last it one day happened; the chair went over and the novelist was on the floor, undismayed, unhurt, and after a moment completing his sentence.

The tremendous effect of Henry James's later novels upon us at Cambridge between the years 1900 and 1905 may be shown by the following facts. During those years I used sometimes to write down, immediately after they had occurred, conversations or scraps of conversation which had seemed to me significant or amusing. I thought

I was recording them verbatim and unembellished, but in fact, as will appear, Henry James—unknown to myself and himself—occasionally took a hand and gave them a perhaps not altogether illegitimate twist. These fragments had, I thought, all disappeared in Ceylon bungalows and in those appalling diaspora of possessions which takes place when one moves from one house to another. But not so very long ago I found in an old notebook two dirty, yellowed, folded sheets of paper—two contemporary records of two such conversations. I propose to give them here.

A word must be said about the dramatis personae in the first conversation. G. was R. K. Gaye, H. was G. H. Hardy, and F. was Walter Morley Fletcher; they were all Fellows of Trinity in 1903 when the conversation took place. Hardy was one of the most strange and charming of men. A "pure" mathematician of the greatest brilliance, he became an F.R.S. and Savilian Professor of Geometry in Oxford. He had the eyes of a slightly startled fawn below the very beautiful and magnificent forehead of an infant prodigy. He gave one the feeling that he belonged more properly to Prospero's island than the Great Court of Trinity. He lived in a double suite of rooms in Great Court with Gaye, a saturnine classical Fellow who committed suicide some years later. Gaye and Hardy were absolutely inseparable; they were never seen apart and rarely talked to other people. They collected railway tickets (this, I think was really Gaye's mania) and had a passion for every kind of game. They admitted Saxon and me into a restricted acquaintanceship because Saxon had been at Westminster with Gaye, and we played bowls with them in the Fellows' Garden and cricket with a walking-stick and tennis ball in their rooms. Fletcher became Secretary of the Medical Research Council, and F.R.S. and a K.B.E. Here is the record of the conversation to which I had given the title, "The Cat, the Worms, and the Rats":

> When I went into G. and H.'s room, I found them sitting one on each side of the fire in a very dejected condition. On the floor between them sat their cat. They were quite silent and dishevelled and they merely gazed at the cat. The cat's ill, said H. at last in a dull voice. It's got worms, at least that's what the vet said —F. told us to go to him.
>
> Poor thing! I felt I had to say, to break the pause.
>
> It hasn't eaten for two days, he went on. You see, the vet says as soon as it makes the movements preparing to eat, the worms—they're in the stomach, you know—come up into the oesophagus and nearly choke it.
>
> But what are you doing for it? I said.
>
> Well, the vet gave us a powder for it. He said just give it to him and it will kill the worms inside. But that's the worst of these experts, you always think it's quite easy when they are telling you what to do; when you go and try to do it, you find it's impossible. We can't get him to take the powder; we tried to make him take it mixed with milk—the vet told us to—but we could only force a little of it down his throat with a teaspoon and even then he was sick at once. F. says he doesn't believe you can make a cat take anything against its will.
>
> There was a long pause, while we all looked at the cat.
>
> It's in a very emaciated condition, H. was pursuing

in a still lower voice, when the door opened and F. came in.

Well, how's the patient? he said with conscientious cheerfulness.

Just the same, said H. You know we can't make it take the powders; it was sick when we forced a little down its throat.

If it was only a dog, I said. You'd simply open its mouth and drop it down. But then of course there are the claws. You can't get hold of a cat.

No, said F., even a dog can't kill a cat easily.

That's because he can't come to close quarters, I said. I suppose if he got it in the back like he does a rabbit, he could quite easily.

I suppose he could, said F. You know a terrier kills a rabbit or rat with a flick just breaking its back. By the bye that reminds me of the most repulsive sight I ever saw—it really was too filthy. It was in France last vac. I was biking in the Rhone valley with J.——

Here H. to whom G. had whispered something broke in: I'm sorry, F., but, before I forget, do you think it's a funny or bad symptom that while the cat is being sick it walks backwards?

Yes, said G., and it also keeps on drawing its head back.

I really don't know, returned F. Well, we were biking through a small village and found there was a fair going on, so we dismounted to have a look. The great attraction was Madame Boug, the champion rat catcher. We found a big crowd awaiting her arrival round a pit. We squeezed in among them and soon she made her appearance. She was a tall, big woman and stark naked except for a tightly fitting red pair of drawers—really quite repulsive, you know. Well, she went into the pit and they loosed about twenty big sewer rats into it too. Then she went down on her hands and knees and chased the rats round. She crawled extraordinarily quickly and every now and then made a dart with her head, caught one by the back in her mouth, gave a little flick, and it was dead. It was quite foul, you know; to see her seize them in her teeth and give that little jerk just like a terrier.

But didn't they bite her? I said.

O yes, he said, in the ears, that was so repulsive. For when there were only three left, she worked them up into a corner and as she was killing one another seized her ear, and I saw another leap up from under her breast right over her neck. I daresay they bit her in the breast too, but it was really so repulsive, you know, that we made off feeling quite sick.

But, said G., as F. got up, I shouldn't have thought her mouth was big enough to seize a rat in.

Ah, said F., she was a big-mouthed woman, quite repulsive, you know.

Then the door closed upon his Goodnight!

May 10th, 1903.

The second conversation, which is a good deal more Jamesian, requires a word of explanation. The dramatis personae are St., Lytton Strachey; The G., Thoby Stephen, who was nicknamed The Goth; S-T, Saxon Sydney-Turner; and M., a rather older scholar called Maclaren. "The method" referred to in the conversation had been invented by Lytton and me; it was a kind of third-degree psychological investigation applied to the souls of one's friends. Though it was a long time before we had any knowledge of Freud, it was a kind of compulsory psycho-analysis. It was intended to reveal to us, and incidentally to the victim, what he was really like; the theory was that by imparting to all concerned the deeper psychological truths, personal relationships would be much improved. Its technique was derived partly from Socrates, partly from Henry James, partly from G. E. Moore, and partly from ourselves. We had already applied the method with disastrous success to Saxon. Here is the conversation:

Sunday night.

I was writing a letter when the St. came in. The G. and S-T. were sitting silent round the fire.

Wait a moment! he said.

There was a long pause while he walked up and down. What have you been doing? I said. Two wonderful conversations, was the answer, and then the pacing began again.

It's reached the ultimate, he said at last.

I looked up. O, it's nothing indecent, he said, and —well, it's the penultimate really.

How *did* you do it?

I simply asked him. It's wonderful if it's true. We were hopeless and the method's smashed.

O, I can't believe that.

Yes, but that's the contortion . . . I can't believe it and he can't make me. There are no ups and downs and there are only a few. And he's——

Here the door opened and M. came in. He stood gaping for two minutes and then joined the other two by falling silent into a chair.

He's going—and that of course will be the ultimate —to give me the names.

God!

He began pacing again.

There's no hypnosis even, he went on. The touching and all that—that's not the important part. You see, I *can't* understand it.

But the questions, I said. How could you ask them?

Well, once I *did* think I was lost. But what's so awful for him, poor thing, is that however much he swears it's true, I can't believe him.

Well, you are cruel. I call it sheer brutality.

He stood and drummed on the table with my pen while I lay back in my chair and looked at him. The group round the fire was still silent. Suddenly he turned to go and as suddenly came back to the table. There's one thing more, he said. This has certainly been the most wonderful of all.

Then the door slammed.

May 10th, 1903.

Let me return for a brief moment to Saxon. I have said that we applied "the method" to him with disastrous success. Lytton and I were very fond of him—we had become intimate friends long before Saxon and I had the double set of rooms in Great Court. But the more intimately we got to know him, the more concerned we became about his psychological state. He seemed, even at the age of twenty, to have deliberately withdrawn himself from life, to protect himself from its impact and from the impact of persons, emotions, and things by spinning around himself an elaborate and ingenious series of cocoons. He was thus in the process of successfully stifling his creativeness, his sensitive and subtle intelligence, his affections. He was, as I have said, a character in a Henry James novel, but he would also have seemed more alive in *Crotchet Castle* or *Nightmare Abbey* than in the Cambridge of 1900. Beneath the facade and the veils one felt that there might be an atrophied Shelley.

Lytton and I decided that we ought to apply "the method" to Saxon, to try to make him tear up and break through the veils into life. One evening after dining in Hall we began to apply the third-degree psychological investigation to him about half-past eight and continued it uninterruptedly until five in the morning; when at last he staggered away to bed, we had successfully uncovered the soul of Saxon, but had disastrously confirmed him in the determination to stifle it in an infinite series of veils. Twenty-five years later, I amused myself by writing "characters" of some of my friends after the manner of La Bruyère. Here is one which was suggested to some extent by recollections of Saxon:

> Aristotle sits in a corner of a room spinning, spinning webs around himself. He has been spinning now for thirty years, so that it is rather difficult to see through the web exactly what he really is, sitting there curled up smoking his pipe in the centre of it. Originally before the webs began, if there was such a time—if indeed he did not begin spinning them in his mother's womb—he must have been charming. He might have been Shelley. He might have dreamed dreams of a queer unsubstantial beauty; the fine temper of his mind might have built a philosophy true and beautiful and

unintelligible; he might have had bright and delicate affections; he might have been happy, he might have been in love. Years ago, I suppose, all this showed more clearly than it does now. For I think that Heracleitus and Aristophanes must have seen it when they took Aristotle to their bosom. It wanted clear eyes to see through the web even then, it wants still clearer eyes now. You go into a large dirty room full of dead things and abominations and uglinesses. The most abominable thing in it are the books; even the *Phaedrus* becomes a degradation there. All the books are dead, and all the thoughts and words of them have become dust and ashes and desolation. You feel that the Rabelais which you had in your overcoat pocket when you came in has already turned into a skeleton of dry bones. There are books everywhere: on tables and chairs and floor and mantelpiece and bed, and scattered among the books are old bottles of medicine and horrible little boxes of tabloids and capsules and pills. You brighten up when you see a copy of the *Lysistrata* lying upon the table; you open it and find a bottle of laudanum between the leaves, thrust in to mark the place. A thin layer of dust and soot lies upon everything. You sink sadly into a chair and look into the corner and there you see an immense accumulated mass of grey strands, dusty, dirty, tangled. They float about the room brushing softly against your face. You shudder? You try to rouse yourself? You talk loud, brutally, not knowing quite what you are saying? Your noise and excitement, my friend, are quite useless; you had much better sit down again and quietly watch him spinning quietly in the corner. Do you see how the web is growing?

There, that long dusty, whitish-grey strand is a list of all the writers on the Higher Mathematics whose names begin with P. A good wrap for the soul? And then there are 124 volumes of Diodorus Siculus and Duns Scotus and Hippocrates and Galen and the Montenegrin poets and the Hottentot philosophers. Fine wraps for the soul? But above all there is the past: to spin the past over the present until what was the present has become the past ready to be spun again over the present that was the future! Quick, let us cover our souls with the litter of memories and old sayings and the dead letters of the dead. And if the dead are ourselves, so much the better; let the rubbish of the past stifle our feelings, let the sap and vigour of our thoughts dry up and ooze away into the dusty accretions which we spin over ourselves. Such is the philosophy of Aristotle. Is he happy? Is the mole or the barnacle or the spider happy? If they are, then Aristotle is too when he has not got the toothache, which is not often. In the very centre of the web, I think, there is still a gentle titillation of unsubstantial happiness whenever he finds another higher mathematician whose name begins with P. or when between 1 and 2 a.m. he explains to Aspasia that the great uncle of his mother's cousin moved in 1882 from Brixton to Balham and that his name was Beeley Tupholme, or even when he sees in his old letters that he was young once with Heracleitus and Aristophanes. It may be that affection still moves him for Aristophanes and Heracleitus and Kyron and Lysistrata and Aspasia, but they move, I think, through the past. The reason of all this? you ask. It may be that God made him—a eunuch; or it may be that the violence and brutality of life were too strong for the delicacy of him; he was terrified by it and by his feelings. He looks sometimes like a little schoolboy whom life has bullied into unconsciousness. Which is really true nobody will ever know, for now he will go on sitting there in his corner spinning his interminable cocoon until he dies. It will be some time before we find out that he really is dead and then we shall go to the large dirty room and push and tear our way through the enormous web which by that time will almost completely fill it, and at last when we stand choking in the centre of it we shall find just nothing at all. Then we shall bury the cocoon.

Lytton Strachey, Thoby Stephen, and Clive Bell all came up to Trinity in the same year as Saxon and I did and we soon got to know them well. We were intimate friends—particularly Lytton, Saxon, and myself—but intimacy in 1900 among middle-class males was different from what it became in generations later than ours. Some of us were called by nicknames; for instance we always called Thoby Stephen The Goth, but we never used christian names. Lytton always called me Woolf and I always called him Strachey until I returned from Ceylon in 1911 and found that the wholesale revolution in society and manners which had taken place in the preceding seven years involved the use of christian names in place of surnames. The difference was—and is—not entirely unimportant. The shade of relationship between Woolf and Strachey is not exactly the same as that between Leonard and Lytton. The surname relationship was determined by and retained that curious formality and reticence which the nineteenth-century public school system insisted upon in certain matters. Now, of course, the use of christian names and their diminutives has become so universal that it may soon perhaps become necessary to indicate intimacy by using surnames.

Lytton was a very strange character already when he came up to Cambridge in 1899. There was a mixture of arrogance and diffidence in him. His mind had already formed in a Voltairean mould, and his inclinations, his passions, the framework of his thought belonged to the eighteenth century, and particularly to eighteenth century France. His body was long, thin, and rather ungainly; all his movements, including his walk, were slow and slightly hesitant—I never remember to have seen him run. When he sat in a chair, he appeared to have tied his body, and particularly his legs, into what I always called a Strachean knot. There was a Strachean voice, common to him and to all his nine brothers and sisters (much less marked in the eldest brother, Dick, who was a major in the army when I knew him, than in the others). It was mainly derived, I think, from the mother and consisted in an unusual stress accent, heavy emphasis on words here and there in a sentence, combined with an unusual tonic accent, so that emphasis and pitch continually changed, often in a kind of syncopated rhythm. It was extremely catching and most people who saw much of Lytton acquired the Strachey voice and never completely lost it. Lytton himself added another peculiarity to the family cadence. Normally his voice was low and fairly deep, but every now and again it went up into a falsetto, almost a squeak.

This squeak added to the effect of his characteristic

style of wit. He was one of the most amusing conversationalists I have ever known. He was not a monologuist or a raconteur. Except when he was with one or a few intimate friends, he did not say very much and his silences were often long. They were often broken by a Strachean witticism, probably a devastating reductio ad absurdum —the wit and the devastation owing much to the perfect turn of the sentence and the delicate stiletto stab of the falsetto voice. Many, particularly among the young, as I said, caught his method of talking and ever afterwards spoke in the Strachey voice; so too, many caught his method of thinking and thought ever after with a squeak in their minds. The unwary stranger, seeing Lytton contortedly collapsed in a tangle of his own arms and legs in the depths of an armchair, his eyes gazing in fixed abstraction through his strong glasses at his toes which had corkscrewed themselves up and round to within a foot of his nose, the unwary stranger might and sometimes did dismiss him as a gentle, inarticulate, nervous, awkward intellectual. All these adjectives were correct, but woe betide the man or woman who thought that they were the end of the matter and of Lytton Strachey. I used to tell him that, when he came to see us and we were not alone, I proposed to put a notice on the arm of his chair: BE CAREFUL, THIS ANIMAL BITES.

The animal bit because, behind the gentleness, the nervousness, and the cynicism, there were very considerable passions. They were the passions of the artist and of the man who is passionately attached to standards of intellectual integrity. This may sound priggish to some people, but no man has ever been less of a prig than he was. He suffered the stupid and stupidity, and the philistine and philistinism, with unconcealed irritation which might take the form either of the blackest, profoundest silence or of a mordant witticism. As he could on occasions be ruthless and inconsiderate, I have known his intolerance produce intolerable situations. One summer he came to stay with us at Rodmell for a few days and, when he heard that a well-known literary man, whom I will call X, had taken a cottage in the village, he asked me to have him in to dinner one evening, as he would like to meet him. I deprecated the idea, as X was, like many literary men, rather a bore and not at all "bright" in the Strachean sense. However Lytton insisted that he had never met X and wanted to meet him, so I foolishly gave way and X appeared the following evening at 7.30. After the first five minutes, Lytton withdrew into himself and a thick cloud of silence, fixing his eyes upon his food or upon the ceiling and tying his legs into even more complicated knots than usual. When X left some three hours later, I do not think that he had heard more than twenty words from the author of *Eminent Victorians*.

This kind of arrogance and rudeness, alternating, as it did, with a curious diffident nervousness, roused a certain amount of hostility to Lytton both among people who knew him a little and often among people who did not know him at all. His physical appearance and voice had that indefinable quality which tends to excite animosity or ridicule at sight in the ordinary man, the Cambridge "blood" or tough, for instance. To his intimate friends, though he could be momentarily infuriating, he was extraordinarily affectionate and lovable. It should be added that many of his characteristics which superficially irritated or repelled people were due to his health. Though he never during the time that I knew him had a dangerous illness before the final cancer which killed him at the age of fifty-one in 1932, I have the impression, looking back over the thirty-one years of my knowing him, that he was hardly ever completely well, or rather that the standard of his physical strength, health, vitality was, compared with the average human being's, low. One felt that he always had to husband his bodily forces in the service of his mind and that, in view of the precarious balance of physical health, it was surprising how much he accomplished.

The characters in *The Waves* are not drawn from life, but there is something of Lytton in Neville. There is no doubt that Percival in that book contains something of Thoby Stephen, Virginia's brother, who died of typhoid, aged twenty-six in 1906. Thoby came up to Trinity from Clifton with an exhibition in the same year as Lytton, Saxon, and I. He gave one an impression of physical magnificence. He was six foot two, broad-shouldered and somewhat heavily made, with a small head set elegantly upon the broad shoulders so that it reminded one of the way in which the small head is set upon the neck of a well-bred Arab horse. His face was extraordinarily beautiful and his character was as beautiful as his face. In his monolithic character, his monolithic common-sense, his monumental judgments he continually reminded one of Dr. Johnson, but a Samuel Johnson who had shed his neuroticism, his irritability, his fears. He had a perfect "natural" style of writing, flexible, lucid, but rather formal, old-fashioned, almost Johnsonian or at any rate eighteenth century. And there was a streak of the same natural style in his talk. Any wild statement, speculative judgment, or Strachean exaggeration would be met with a "Nonsense, my good fellow", from Thoby, and then a sentence of profound, but humorous, common-sense, and a delighted chuckle. Thoby had a good sense of humour, a fine, sound, but not brilliant mind. He had many of the characteristic qualities of the males of his family, of his father Leslie Stephen, his uncle James Fitzjames Stephen and his cousin J. K. Stephen. But what everyone who knew him remembers most vividly in him was his extraordinary charm. He had greater personal charm than anyone I have ever known, and, unlike all other great "charmers", he seemed, and I believe was, entirely unconscious of it. It was, no doubt, partly physical, partly due to the unusual combination of sweetness of nature and affection with rugged intelligence and a complete lack of sentimentality, and partly to those personal flavours of the soul which are as unanalysable and indescribable as the scents of flowers or the overtones in a line of great poetry.

Thoby was an intellectual; he liked an argument and had a great, though conservative and classical, apprecia-tion and love of literature. But he also, though rather scorn-ful of games and athletics, loved the open air—watching birds, walking, following the beagles. In these occupa-tions, particularly in walking, I often joined him. Walking with him was by no means a tame business, for it was almost a Stephen principle in walking to avoid all roads and ig-nore the rights of property owners and the law of trespass. Owing to these principles we did not endear ourselves to the gamekeepers round Cambridge. Though fundament-ally respectable, conservative, and a moralist, he was al-ways ready in the country to leave the beaten track in more senses than one. In our walks up the river towards Trumpington, we had several times noticed a clump of magnificent hawthorn trees in which vast numbers of starlings came nightly to roost. I have never seen such enormous numbers of birds in so small a space; there must have been thousands upon thousands and the trees were in the evening literally black with them. We several times tried to put them all up into the air at the same time, for, if we succeeded, it would have been a marvellous sight to see the sky darkened and the setting sun obscured by the immense cloud of birds. But we failed because every time we approached the trees, the birds went up into the air spasmodically in gusts, and not altogether. So we bought a rocket and late one evening fired it from a distance into the trees. The experiment succeeded and we had the pleasure of seeing the sun completely blotted out by starlings. It was several years later that I was to see as large or even larger flights of birds in Ceylon—the great flocks of teal wheeling round the lagoons or the tanks in the Hambantota district.

The following letter which Thoby wrote to me shortly after I went to Ceylon gives, I think, some faint flavour of his character:

46 Gordon Square,
Bloomsbury.
Jan 15, 1905.

Dear Woolf,

I ought to have written to you before this, but the world has been very barren of circumstance, and one feels that a letter ought to contain information in some proportion to the number of miles it has to go. That is probably a fallacy, but anyhow you must have had incidents enough by now among the Obesekeridae to fill an epistle which I hope this will evoke. I have been plodding pretty steadily at the law and becoming crystallised at it—in fact my moustache has disappeared. I was in the New Forest at Christmas, where I got some hunting, one especially rare chivy. From there I walked to Hindhead and stayed some days with Pollock. There were there J. Pollock, his sister and her husband Waterlow (you know the man I suppose), Meredith for a time, and old Bell. I more or less enjoyed it but it was damned funny. Waterlow is a serious cove and devilish Cambridge. "What is poetry? Well, there you ask me a difficult question—I am not sure that it *is* anything—it depends what you mean by being" and so on the old round, till after an hour or two all go to bed leaving Bell and me who shout simultaneously "Now let's talk about hunting." His wife lags behind him but struggles gallantly "Sidney, Sidney, what do you mean by Mon-og-amy?" However he has a bottom of good sense and is not a bad fellow . . . The good old chapel row[1] is still fermenting. Cornford and Gaye have pamphleteered and Pollock is following with a "legal aspect" one. I have suppressed mine pro tem out of deference to Cornford—who takes I think a rotten line—chapel is either the sublimest function of man or the most pathetic of human fallacies—it's no good being dainty with Christians and chapel's obviously rot and nothing else. I seem to have done nothing and seen nobody and read little of interest for the deuce of a time —I've been reading satires chiefly when I've had time— I think probably all the best things written have been

[1] Thoby had written and circulated in Cambridge a pamphlet against the practice of compulsory chapel in Trinity College.

satires except Virgil—and one can worm a quasi-satire out of the bees. I think I am going to make my working men[1] read it. Virgil after all is the top of the tree and Sophocles is thereabouts—next come Catullus and Aristophanes, that is my mature opinion so far as the ancients go. They talk of abolishing Greek at Cambridge and Jackson and Verrall are helping the devils. If they do you'd better become a naturalized Cingalese—and I shall go to the Laccadives. Haynes annoys me rather—I dined with Bell and him the other day—he talks of nothing but suicides, disease and bawdy, and his beastly book—almost he persuades me to be a Christian. Well, my good fellow, I've nothing to say but what's unutterably dull, but I hope I shall hear something enlivening from you some day.

Yrs. ever
J. T. Stephen.

Clive Bell came up to Trinity the same year as we did, 1899, and when we first got to know him he was different in many ways from us and even from the Clive Bell whom I found married to Vanessa Stephen and living in Gordon Square when I returned from Ceylon in 1911. He came into our lives because he got to know Saxon, having rooms on the same staircase in New Court. Lytton, Saxon, Thoby, and I belonged, unconcealably and unashamedly, to that class of human beings which is regarded with deep suspicion in Britain, and particularly in public schools and universities, the intellectual. Clive, when he came up to Trinity from Marlborough, was not yet an intellectual. He was superficially a "blood". The first time I ever saw

[1] Thoby taught at the Working Men's College.

him he was walking through Great Court in full hunting rig-out, including—unless this is wishful imagination—a hunting horn and the whip carried by the whipper-in. He was a great horseman and a first-rate shot, very well-off, and to be seen in the company of "bloods", not the rowing, cricket, and rugger blues, but the rich young men who shot, and hunted, and rode in the point-to-point races. He had a very attractive face, particularly to women, boyish, goodhumoured, hair red and curly, and what in the eighteenth century was called, I think, a sanguine complexion.

Clive became great friends with Thoby, for they both were fond of riding and hunting. In those early days, and indeed for many years afterwards, intellectually Clive sat at the feet of Lytton and Thoby. He was one of those strange Englishmen who break away from their environment and become devoted to art and letters. His family of wealthy philistines, whose money came from coal, lived in a large house in Wiltshire. Somehow or other, Mr. and Mrs. Bell produced Clive's mind which was a contradiction in terms of theirs.[1] For his mind was eager, lively, intensely curious, and he quickly developed a passion for literature[2] and argument. We had started some reading

[1] I do not think that I ever met either of Clive's parents, but I have heard so much about them from him and from others that I have no doubt about the truth of what I say here.

[2] It is worth noting that in those days we set little or no store by pictures and painting. I never heard Clive talk about pictures at Cambridge, and it was only after he came down and lived for a time in Paris and got to know Roger Fry that his interest in art developed. Music already meant a good deal to Lytton, Saxon, and me and we went to chamber music concerts in Cambridge and orchestral concerts in London, but I do not think that it has ever meant much to Clive.

societies for reading aloud plays, one of which met at midnight, and Clive became a member of them. In this way we came to see a good deal of him and his admiration for Lytton and Thoby began to flourish.

It is necessary here to say something about the Society—The Apostles—because of the immense importance it had for us, its influence upon our minds, our friendships, our lives. The Society was and still is "secret", but, as it has existed for 130 years or more, in autobiographies and biographies of members its nature, influence, and membership have naturally from time to time been described. There is a good deal about it in the autobiography of Dean Merivale, who was elected in 1830, and in the memoir of Henry Sidgwick, who was elected in 1856, and information about its condition in the early years of the present century can be found in *The Life of John Maynard Keynes* by R. F. Harrod, who was not himself an Apostle. These descriptions show that its nature and atmosphere have remained fundamentally unaltered throughout its existence. The following words from Sidgwick's *A Memoir* are worth quoting:

> I became a member of a discussion society—old and possessing historical traditions—which went by the name of "The Apostles". When I joined it the number of members was not large, and there is an exuberant vitality in Merivale's description to which I recall nothing corresponding. But the spirit, I think, remained the same, and gradually this spirit—at least as I apprehended it—absorbed and dominated me. I can only describe it as the spirit of the pursuit of truth with absolute devotion and unreserve by a group of intimate

> friends, who were perfectly frank with each other, and indulged in any amount of humorous sarcasm and playful banter, and yet each respects the other, and when he discourses tries to learn from him and see what he sees. Absolute candour was the only duty that the tradition of the society enforced . . . It was rather a point of the apostolic mind to understand how much suggestion and instruction may be derived from what is in form a jest—even in dealing with the gravest matters . . . It came to seem to me that no part of my life at Cambridge was so real to me as the Saturday evenings on which the apostolic debates were held; and the tie of attachment to the society is much the strongest corporate bond which I have known in life.

The Apostles of my generation would all have agreed with every word in this quotation. When Lytton, Saxon, and I were elected, the other active undergraduate members were A. R. Ainsworth, Ralph Hawtrey, and J. T. Sheppard.[1] When Maynard Keynes came up, we elected him in 1903. Sidgwick says that the Society absorbed and dominated him, but that is not quite the end of the story. Throughout its history, every now and again an Apostle has dominated and left his impression, within its spirit and tradition, upon the Society. Sidgwick himself was one of these, and a century ago he dominated the Society, refertilizing and revivifying its spirit and tradition. And what Sidgwick did in the fifties of last century, G. E. Moore was doing when I was elected.

[1] Ainsworth became a civil servant in the Education Office; Hawtrey, now Sir Ralph Hawtrey, a civil servant in the Treasury; Sheppard, now Sir J. T. Sheppard, Provost of King's College, Cambridge.

Mrs. Sidney Webb once said to me: "I have known most of the distinguished men of my time, but I have never yet met a great man." I had admiration and affection for Beatrice Webb, but when, in her cold and beautiful voice, she pronounced one of these inexorable Sinaic judgments in her tenebrous Grosvenor Road dining-room, gazing through the window across the river at the Doulton China Works, I used to feel that in one moment I should be submerged in despair and desolation, that I was a miserable fly crawling painfully up the Webb's window to be swatted, long before I reached the top, by their merciless common-sense. But sometimes the fly gave a dying kick, and on this occasion I said: "I suppose you don't know G. E. Moore." No, she said, she did not know G. E. Moore, though she knew, of course, whom I meant, and the question of human greatness having been settled, we passed to another question.

The author of Ecclesiasticus probably agreed with Beatrice Webb, for he asked us to praise not great men but famous men—a very different thing. The conversation in Grosvenor Road took place forty years ago, but I still think despite the two impressive authorities that I was right, that George Moore was a great man, the only great man whom I have ever met or known in the world of ordinary, real life. There was in him an element which can, I think, be accurately called greatness, a combination of mind and character and behaviour, of thought and feeling which made him qualitatively different from anyone else I have ever known. I recognize it in only one or two of the many famous dead men whom Ecclesiasticus and others enjoin us to praise for one reason or another.

It was, I suppose, in 1902 that I got to know Moore

well. He was seven years my senior and already a Fellow of Trinity. His mind was an extraordinarily powerful instrument; it was Socratic, analytic. But unlike so many analytic philosophers, he never analyzed just for the pleasure or sake of analysis. He never indulged in logic-chopping or truth-chopping. He had a passion for truth, but not for all or any truth, only for important truths. He had no use for truths which Browning called "dead from the waist down". Towards the end of the nineteenth century there was an extraordinary outburst of philosophical brilliance in Cambridge. In 1902, among the Fellows of Trinity were four philosophers, all of whom were Apostles: J. E. McTaggart, A. N. Whitehead, Bertrand Russell, and G. E. Moore. McTaggart was one of the strangest of men, an eccentric with a powerful mind which, when I knew him, seemed to have entirely left the earth for the inextricably complicated cobwebs and *O altitudos* of Hegelianism. He had the most astonishing capacity for profound silence that I have ever known. He lived out of college, but he had an "evening" once a week on Thursdays when, if invited or taken by an invitee, you could go and see him in his rooms in Great Court. The chosen were very few, and Lytton, Saxon, and I, who were among them, every now and again nerved ourselves to the ordeal. McTaggart always seemed glad to see us, but, having said good evening, he lay back on a sofa, his eyes fixed on the ceiling, in profound silence. Every five minutes he would roll his head from side to side, stare with his rather protuberant, rolling eyes round the circle of visitors, and then relapse into immobility. One of us would occasionally manage to think of something banal and halting to say, but I doubt whether I ever heard McTaggart initiate a conversation, and when he did say something it was usually calculated to bring to a sudden end any conversation initiated by one of us. Yet he did not seem to wish us not to be there; indeed, he appeared to be quite content that we should come and see him and sit for an hour in silence.

In the early 1890's McTaggart's influence was great. He was six years older than Russell and seven years older than Moore, and these two in their early days at Trinity were first converted to Hegelianism by McTaggart. But Moore could never tolerate anything but truth, commonsense, and reality, and he very soon revolted against Hegel: Bertrand Russell describes the revolt in the following words:

> Moore, first, and I closely following him, climbed out of this mental prison and found ourselves again at liberty to breathe the free air of a universe restored to reality.

When I came up to Trinity, McTaggart, though regarded with respect and amused affection as an eccentric, had completely lost his intellectual and philosophical influence. The three other philosophers' reputation was great and growing, and they dominated the younger generation. In 1902 Whitehead was forty-one years old, Russell thirty, and Moore twenty-nine. It is a remarkable fact—a fine example of our inflexible irrationality and inveterate inconsistency—that, although no people has ever despised, distrusted and rejected the intellect and intellectuals more than the British, these three philosophers were each awarded the highest and rarest of official honours, the Order of Merit. 1903 was an *annus mirabilis* for Cambridge philosophy, for in that year were published Russell's *Principles of Mathematics* and Moore's *Principia Ethica*. Russell used to come to Moore's rooms sometimes in order to discuss some difficult problem that was holding him up. The contrast between the two men and the two minds was astonishing and fascinating. Russell has the quickest mind of anyone I have ever known; like the greatest of chess players he sees in a flash six moves ahead of the ordinary player and one move ahead of all the other Grand Masters. However serious he may be, his conversation scintillates with wit and a kind of puckish humour flickers through his thought. Like most people who possess this kind of mental brilliance, in an argument a slower and duller opponent may ruefully find that Russell is not always entirely scrupulous in taking advantage of his superior skill in the use of weapons. Moore was the exact opposite, and to listen to an argument between the two was like watching a race between the hare and the tortoise. Quite often the tortoise won—and that, of course, was why Russell's thought had been so deeply influenced by Moore and why he still came to Moore's rooms to discuss difficult problems.

Moore was not witty; I do not think that I ever heard him say a witty thing; there was no scintillation in his conversation or in his thought. But he had an extraordinary profundity and clarity of thought, and he pursued truth with the tenacity of a bulldog and the integrity of a saint. And he had two other very rare characteristics. He had a genius for seeing what was important and what was unimportant and irrelevant, in thought and in life and in persons, and in the most complicated argument or situation he pursued the relevant and ignored the irrelevant with amazing tenacity. He was able to do so because of the second characteristic, the passion for truth (and, as I shall show, for other things) which burned in him. The tortoise so often won the race because of this combination of clarity, integrity, tenacity, and passion.

The intensity of Moore's passion for truth was an integral part of his greatness, and purity of passion was an integral part of his whole character. On the surface and until you got to know him intimately he appeared to be a very shy, reserved man, and on all occasions and in any company he might fall into profound and protracted silence. When I first got to know him, the immensely high standards of thought and conduct which he seemed silently to demand of an intimate, the feeling that one should not say anything unless the thing was both true and worth saying, the silences which would certainly envelope him and you, tinged one's wish to see him with some anxiety, and I know that standing at the door of his room, before knocking and going in, I often took a deep breath just as one does on a cool day before one dives into the cold green sea. For a young man it was a formidable, an alarming experience, but, like the plunge into the cold sea, once one had nerved oneself to take it, extraordinarily exhilarating. This kind of tension relaxed under the influence of time, intimacy, and affection, but I do not think that it ever entirely disappeared—a proof, perhaps, of the quality of greatness which distinguished Moore from other people.

His reserve and silences covered deep feeling. When Moore said: "I *simply* don't understand *what* he means," the emphasis on the "simply" and the "what" and the shake of his head over each word gave one a glimpse of the passionate distress which muddled thinking aroused in him. We used to watch with amusement and admiration the signs of the same thing when he sat reading a book, pencil in hand, and continually scoring the poor wretch of a writer's muddled sentences with passionate underlinings and exclamation marks. I used to play fives with him at Cambridge, and he played the game with the same passion as that with which he pursued truth; after a few minutes in the court the sweat poured down his face in streams and soaked his clothes—it was excitement as well as exercise. It was the same with music. He played the piano and sang, often to Lytton Strachey and me in his rooms and on reading parties in Cornwall. He was not a highly skilful pianist or singer, but I have never been given greater pleasure from playing or singing. This was due partly to the quality of his voice, but principally to the intelligence of his understanding and to the subtlety and intensity of his feeling. He played the Waldstein sonata or sang "Ich grolle nicht" with the same passion with which he pursued truth; when the last note died away, he would sit absolutely still, his hands resting on the keys, and the sweat streaming down his face.

Moore's mind was, as I said, Socratic. His character, too, and his influence upon us as young men at Cambridge were Socratic. It is clear from Plato and Xenophon that Socrates's strange simplicity and integrity were enormously attractive to the young Athenians who became his disciples, and he inspired great affection as well as admiration. So did Moore. Plato in the *Symposium* shows us a kind of cosmic absurdity in the monumental simplicity of Socrates; and such different people as Alcibiades, Aristophanes, and Agathon "rag" him about it and laugh at him gently and affectionately. There was the same kind of divine absurdity in Moore. Socrates had the great advantage of combining a very beautiful soul with a very ugly face, and the Athenians of the fifth century B.C. were just the people to appreciate the joke of that. Moore had not that advantage. When I first knew him, his face was amazingly beautiful, almost ethereal, and, as Bertrand Russell has said, "he had, what he retained throughout his life, an extraordinarily lovable smile". But he resembled Socrates in possessing a profound simplicity, a simplicity which Tolstoy and some other Russian writers consider to produce the finest human beings. These human beings are "simples" or even "sillies"; they are absurd in ordinary life and by the standards of sensible and practical men. There is a superb description of a "silly" in Tolstoy's autobiography and, of course, in Dostoevsky's *The Idiot*. In many ways Moore was one of these divine "sillies". It showed itself perhaps in such simple, unrestrained, passionate gestures as when, if told something particularly astonishing or confronted by some absurd statement at the crisis of an argument, his eyes would open wide, his eyebrows shoot up, and his tongue shoot out of his mouth. And Bertrand Russell has described the pleasure with which one used to watch Moore trying unsuccessfully to light his pipe when he was arguing an important point. He would light a match, hold it over the bowl of his pipe until it burnt his fingers and he had to throw it away, and go on doing this—talking the whole time or listening intently to the other man's argument—until the whole box of matches was exhausted.

After I went to Ceylon, Moore wrote me some letters

which I think, may give to those who never knew him some feeling of his bleak simplicity which was at the same time so endearing, his complete inability to say anything which he did not think or feel, and the psychological atmosphere which surrounded him and which, as I have said, it was both alarming and exhilarating to plunge into. Here are two of his letters:

11 Buccleuch Place
Edinburgh.
March 16, '05

Dear Woolf,

It is very shameful of me not to have answered your letter sooner: it is now nearly three weeks since I got it. I was very, very pleased to get it: I had been wanting to write to you before it came, and was afraid you were not going to write. The reason why I haven't written, in spite of wanting to, is that I don't know what to say. I have begun three letters to you already before this one; but I wouldn't finish them, because they were so bad. I'm afraid I have nothing to say, which is worth saying; or, if I have, I cannot express it.

I do not work any better than I used to, and am just as little interested in anything. As for my work, I have not yet written my review of the "Principles of Mathematics"; and it seems as if I never should. I think I should scarcely get on any faster with my book, even if I had begun that.

Ainsworth and I hardly ever see anyone here. He works very well; and the rest of the time I play the piano to him, or we read together. We have read three of Jane Austen's this term. We have only played golf once a week so far; but we have just been elected members of the club, so that we shall probably play oftener. I think I have improved a little; but I am still very bad indeed.

I went to Cambridge and back a fortnight ago, all within twenty-four hours, to vote in favour of Compulsory Greek. As it turned out, I needn't have gone, for we were 1,500 to 1,000. I found the new brother at Strachey's; but he hardly said anything while I was there, and was not so very attractive at first sight, as I had thought he would be from their description of him.

There is to be a reading-party at Easter; but I almost wish there were not to be: I do feel so incapable of enjoying anything. I wish you could be there. As it is, I don't know who there will be, except Strachey and MacCarthy and ourselves.

My brother[1] has just published a book on Dürer. There is a great deal of philosophy in it, which begins with this sentence: "I conceive the human reason to be the antagonist of all forces other than itself." I do wish people wouldn't write such silly things—things, which, one would have thought, it is so perfectly easy to see to be just false. I suppose my brother's philosophy may have some merits: but it seems to me just like all wretched philosophy—vague, and obviously inconsequent, and full of falsehoods. I think its object is to be like a sermon—to make you appreciate good things; and I sometimes wonder whether it is possible to do this without saying what is false. But it does annoy me terribly that people should admire such things—as they

[1] Thomas Sturge Moore, the poet.

do. I hope you will soon get away from Jaffna. I suppose it is very flat and very hot; and I think there must be more nice Englishmen somewhere in Ceylon.

Ainsworth sends his love.

Yours

G. E. Moore.

6 Pembroke Villas, The Green,
Richmond, Surrey.
December 1, '08.

Dear Woolf,

I am sending you, enclosed with this, a copy of my last publication. I am sending it, because Strachey said he thought you would like to have it, and at the same time gave me your address. Did I send you the one before, published in 1906? That was a much more interesting and important one. I expect I didn't send it; but I can't feel quite sure that I didn't. If I didn't, and you would like it, I will send you that too. I have always remembered very well that I promised, before you went away, to send you everything I published. But it is so difficult, after such a long time, to feel sure that anyone wants such a promise kept.

I wish I had written to you sometimes: that is to say, if you would have liked to hear from me. The last letter I had from you was in May 1905, nearly four years ago; and I never answered it. I did, in fact, write an answer to it once; but I was ashamed to send it. The truth is I cannot write decently to anyone; I always say such silly things, which don't seem to express what I mean. You said in your last letter to me that you felt something like this too; so perhaps you will understand.

I expect Strachey will have told you about our leaving Edinburgh, and about Ainsworth's marrying my sister. I have not found that it makes much difference to me, my coming here. I had hoped that I should see a great deal of people that I wanted to see, and that it would be more like Cambridge again. But in fact I hardly ever see anyone except my family.

I was a good deal excited about the marriage, especially when they were first engaged. You know they were only engaged for two months before they married, and of course Ainsworth was constantly coming here. He was very happy; and I wished I could have been engaged too. You know they are coming to live here with me after Christmas. I suppose it is rather an unusual arrangement; but I don't see why it shouldn't succeed.

I don't think you would find me at all altered from what I was. The only difference I notice is that I seem to find it more and more difficult to write anything about philosophy. The only things I have written in all these years, besides a few reviews in the International Journal and some MSS for private persons, are these two papers—the one I'm sending and the one of 1906. I haven't written a line of my book. I have always been engaged either on reviews or on something else, spending no end of time on them with hardly any result. *All* this year I have been trying to write an article on Hume for MacCarthy's New Quarterly, and nothing is done yet, though I've begun it over and over again. I feel very different about it at different times. Sometimes

I seem to see how I could do it; very often I feel as if I can't or won't try; and, when I do try, I almost always seem to lose the thread: and there are many other different states of mind too.

Are you coming home this next year? I hope, if you do come, I shall see you a great deal. I have heard very little about you.

Yours very affectionately
G. E. Moore.

This single-minded simplicity permeated his life, and the absurdity which it often produced in everyday life added to one's admiration and affection for him. Like Socrates, he attracted a number of friends and followers as different from one another as Plato and Aristophanes were from Alcibiades and Xenophon. They ranged from Lytton Strachey and Desmond MacCarthy to Sir Ralph Wedgwood[1], Lord Keynes, and Sir Edward Marsh.[2] Everyone enjoyed Moore's absurdities, laughed at them, and he shared the enjoyment. For although not himself actively

[1] Ralph Wedgwood was a contemporary of Moore's at Trinity. He was Chief General Manager of London and North-Eastern Railway from 1923 to 1939 and Chairman, Railway Executive Committee from 1939 to 1941. Knighted in 1924, he was created Baronet in 1942.

[2] Eddie Marsh was also a contemporary of Moore's at Trinity. He entered the Civil Service in the Colonial Office in 1896. He was for many years Private Secretary to Winston Churchill, but during his official career he was also Private Secretary to Joseph Chamberlain, Alfred Lyttleton, Asquith, J. H. Thomas, and Malcolm Macdonald. He was a well-known patron of the arts and a man about town with an eye-glass in his eye. But when I was up at Trinity, he still from time to time came up for a weekend and appeared in Moore's rooms.

witty or humorous, he had a fine, sensitive sense of humour. In conversation Lytton Strachey's snake-like witticisms greatly amused him, but the wit and humour which he liked best, I think, were Desmond MacCarthy's. Desmond was half Irish and his humour had the soft, lovely charm which traditionally is characteristic of Ireland. He was a brilliant talker and raconteur, and he could make Moore laugh as no one else could. And Moore laughed, when he did laugh, with the same passion with which he pursued truth or played a Beethoven sonata. A frequent scene which I like to look back upon is Desmond standing in front of a fire-place telling a long, fantastic story in his gentle voice and Moore lying back on a sofa or deep in an armchair, his pipe as usual out, shaking from head to foot in a long paroxysm of laughter.

During the Easter vac. Moore always arranged a "reading party". Only Apostles were asked and only those with whom Moore felt completely at ease, and that by no means covered everyone even among the elect. Of the undergraduates still up at Cambridge Ainsworth, Lytton Strachey, and I used to go and a certain number of older men who had gone down came, at any rate for the Easter holidays. Desmond MacCarthy, Theodore Llewelyn Davies of the Treasury, Robin Mayor of the Education Office, Charlie Sanger, a barrister, and Bob Trevelyan, the poet, were always asked and always came if they could. We went, twice I think, to the Lizard in Cornwall and once to Hunter's Inn in Devon. I enjoyed these "reading parties" enormously; I suppose we did sometimes read something, but in memory the days seem to me to have passed in talking and walking—and if they are good, few things can be better than walking and talking. Both were,

I think, really very good and it was very exciting for a young undergraduate to be able to form intimate friendships with these older men—friendships which in fact lasted until death ended them.[1] In the evenings Moore sang and played for us, and then we talked and argued again. Moore was at his best on these "parties"; he liked everyone and was at his ease with them.

I feel that I must now face the difficult task of saying something about Moore's influence upon my generation. There is no doubt that it was immense. Maynard Keynes in his *Two Memoirs* wrote a fascinating, an extremely amusing account or analysis of this influence of Moore upon us as young men. Much of what he says is, of course, true and biographically or autobiographically important. Maynard's mind was incredibly quick and supple, imaginative and restless; he was always thinking new and original thoughts, particularly in the field of events and human behaviour and in the reaction between events and men's actions. He had the very rare gift of being as brilliant and effective in practice as he was in theory, so that he could outwit a banker, business man, or Prime Minister as quickly and gracefully as he could demolish a philosopher or crush an economist. It was these gifts which enabled him to revolutionize economic theory and national economic and financial policy and practice, and to make a considerable fortune by speculation and a considerable figure in the City and in the world which is concerned with the patronage or production of the arts, and particularly the theatre and ballet. But most people who knew

[1] Today all are dead. Theodore Davies, a man of extraordinary brilliance and great charm, was killed young in a tragic accident. All the others remained intimate friends for thirty or forty years.

him intimately and his mind in shirtsleeves rather than public uniform would agree that there were in him some streaks of intellectual wilfulness and arrogance which often led him into surprisingly wrong and perverse judgments. To his friends he was a lovable character and these faults or idiosyncracies were observed and discounted with affectionate amusement.

It is always dangerous to speak the truth about one's most intimate friends, because the truth and motives for telling it are almost invariably misunderstood. In all the years that I knew Maynard and in all the many relations, of intimacy and business which I had with him, I never had even the ghost of a quarrel or the shadow of unpleasantness, though we often disagreed about things, persons, or policies. He was essentially a lovable person. But to people who were not his friends, to subordinates and to fools in their infinite variety whom one has to deal with in business or just daily life, he could be anything but lovable; he might, at any moment and sometimes quite unjustifiably, annihilate some unfortunate with ruthless rudeness. I once heard him snap out to an auditor who was trying to explain to the Board of Directors of a company some item in the audited accounts: "We all know, Mr. X., that auditors consider that the object of accounts is to conceal the truth, but surely not even you can believe that their object is to conceal the truth from the Directors."

It was this streak of impatience and wilfulness combined with a restless and almost fantastic imagination, which often induced Maynard to make absurdly wrong judgments. But once having committed himself to one of his opinions or judgments, theories or fantasies, he would without compunction use all the powers and brilliance of his mind, his devastating wit and quickness, to defend it, and in the end would often succeed in convincing not only his opponent, but himself. In several points in *Two Memoirs* his recollection and interpretation are quite wrong about Moore's influence, I think. His main point in the memoir is that Moore in *Principia Ethica* propounded both a religion and a system of morals and that we as young men accepted the religion, but discarded the morals. He defines "religion" to mean one's attitude towards oneself and the ultimate, and "morals" to mean one's attitude towards the outside world and the intermediate. Moore's religion which we accepted, according to Maynard, maintained that "nothing mattered except states of mind, our own and other people's of course, but chiefly our own. These states of mind were not associated with action or achievement or with consequences. They consisted in timeless, passionate states of contemplation and communion, largely unattached to 'before' and 'after' . . . The appropriate objects of passionate contemplation and communion were a beloved person, beauty and truth, and one's prime objects in life were love, the creation and enjoyment of aesthetic experience and the pursuit of knowledge. Of these love came a long way first."

Although Maynard calls this doctrine which we accepted a "faith" and a "religion", he says that Moore's disciples and indeed Moore himself regarded it as entirely rational and scientific, and applied an extravagantly rationalistic, scholastic method for ascertaining what states of affairs were or were not good. The resulting beliefs were fantastically idealistic and remote from reality and "real" life. The effect of this curious amalgam of extreme rationalism, unworldliness, and dogmatic belief was intensified by our complete neglect of Moore's "morals". We paid no attention at all to his doctrine of the importance of rightness and wrongness as an attribute of actions or to the whole question of the justification of general rules of conduct. The result was that we assumed that human beings were all rational, but we were complete "immoralists", recognizing "no moral obligation on us, no inner sanction, to conform or obey."

In my recollection this is a distorted picture of Moore's beliefs and doctrine at the time of the publication of his *Principia Ethica* and of the influence of his philosophy and character upon us when we were young men up at Cambridge in the years 1901 to 1904. The tremendous influence of Moore and his book upon us came from the fact that they suddenly removed from our eyes an obscuring accumulation of scales, cobwebs, and curtains, revealing for the first time to us, so it seemed, the nature of truth and reality, of good and evil and character and conduct, substituting for the religious and philosophical nightmares, delusions, hallucinations, in which Jehovah, Christ, and St. Paul, Plato, Kant, and Hegel had entangled us, the fresh air and pure light of plain commonsense.

It was this clarity, freshness, and common-sense which primarily appealed to us. Here was a profound philosopher who did not require us to accept any "religious" faith or intricate, if not unintelligible, intellectual gymnastics of a Platonic, Aristotelian, Kantian, or Hegelian nature; all he asked us to do was to make quite certain that we knew what we meant when we made a statement and to analyze and examine our beliefs in the light of common-sense. Philosophically what, as intelligent young men, we wanted to know was the basis, if any, for our or any scale of values and rules of conduct, what justification there was for our belief that friendship or works of art for instance were good or for the belief that one ought to do some things and not do others. Moore's distinction between things good in themselves or as ends and things good merely as means, his passionate search for truth in his attempt in *Principia Ethica* to determine what things are good in themselves, answered our questions, not with the religious voice of Jehovah from Mount Sinai or Jesus with his sermon from the Mount, but with the more divine voice of plain common-sense.

On one side of us, we were in 1901 very serious young men. We were sceptics in search of truth and ethical truth. Moore, so we thought, gave us a scientific basis for believing that some things were good in themselves. But we were not "immoralists"; it is not true that we recognized "no moral obligation on us, no inner sanction, to conform or obey" or that we neglected all that Moore said about "morals" and rules of conduct. It is true that younger generations, like their elders, were much less politically and socially conscious in the years before the 1914 war than they have been ever since. Bitter experience has taught the world, including the young, the importance of codes of conduct and morals and "practical politics". But Moore himself was continually exercised by the problems of goods and bads as means of morality and rules of conduct and therefore of the life of action as opposed to the life of contemplation. He and we were fascinated by questions of what was right and wrong, what one *ought* to do. We followed him closely in this as in other parts of his doctrine and argued interminably about the consequences of one's actions, both in actual and imaginary situations. Indeed one of the problems which worried us was what part Moore (and we, his disciples) *ought* to play in ordinary life, what, for instance, our attitude *ought* to be towards practical politics. I still possess a paper which I wrote for discussion in 1903 and which is explicitly concerned with these problems. It asks the question whether we ought to follow the example of George Trevelyan,[1] and take part in practical politics, going down into the gloomy Platonic cave, where "men sit bound prisoners guessing at the shadows of reality and boasting that they have found truth", or whether we should imitate George Moore, who though "he has no small knowledge of the cave dwellers, leaves alone their struggles and competitions." I said that the main question I wanted to ask was: "Can we and ought we to combine the two Georges in our own lives?" And was it rational that George Moore, the philosopher, should take no part in practical politics, or "right that we should as we do so absolutely ignore their questions?" My answer in 1903 was perfectly definite that we *ought* to take part in practical politics and the last words of my paper are: "While philosophers sit outside the cave, their philosophy will never reach politicians or people, so that after all, to put it plainly, I *do* want Moore to draft an Education Bill."

[1] George Macaulay Trevelyan, O.M., the historian and for many years Master of Trinity College. He was four years my senior at Trinity and, when I first knew him, had just become a Fellow of the college. He was a rather fiercely political young man.

I have said that we were very serious young men. We were, indeed, but superficially we often appeared to be the exact opposite and so enraged or even horrified a good many people. After all we were young once—we were young in 1903; and we were not nearly as serious and solemn as we appeared to some people. We were serious about what we considered to be serious in the universe or in man and his life, but we had a sense of humour and we felt that it was not necessary to be solemn, because one was serious, and that there are practically no questions or situations in which intelligent laughter may not be healthily catalytic. Henry Sidgwick, in his *Memoir*, looking back in old age to the year 1856 when he was elected an Apostle wrote:

> No consistency was demanded with opinions previously held—truth as we saw it then and there was what we had to embrace and maintain, and there were no propositions so well established that an Apostle had not the right to deny or question, if he did so sincerely and not from mere love of paradox. The gravest subjects were continually debated, but gravity of treatment, as I have said, was not imposed, though sincerity was. In fact it was rather a point of the apostolic mind to understand how much suggestion and instruction may be derived from what is in form a jest—even in dealing with the gravest matters.

I am writing today just over a century after the year in which Sidgwick was elected an Apostle, and looking back to the year 1903 I can say that our beliefs, our discussions, our intellectual behaviour in 1903 were in every conceivable way exactly the same as those described by Sidgwick. The beliefs "fantastically idealistic and remote from reality and real life", the absurd arguments, "the extravagantly scholastic" method were not as simple or silly as they seemed. Lytton Strachey's mind was fundamentally and habitually ribald and he had developed a protective intellectual facade in which a highly personal and cynical wit and humour played an important part. It was very rarely safe to accept the face value of what he said; within he was intensely serious about what he thought important, but on the surface his method was to rely on "suggestion and instruction derived from what is in form a jest—even in dealing with the gravest matters." I think that in my case, too, there was a natural tendency to express myself ironically—and precisely in matters or over questions about which one felt deeply as being of great importance—for irony and the jest are used, particularly when one is young, as antidotes to pomposity. Of course we were young once; we were young in 1903, and we had the arrogance and the extravagance natural to the young.

The intellectual, when young, has always been in all ages enthusiastic and passionate and therefore he has tended to be intellectually arrogant and ruthless. Our youth, the years of my generation at Cambridge, coincided with the end and the beginning of a century which was also the end of one era and the beginning of another. When in the grim, grey, rainy January days of 1901 Queen Victoria lay dying, we already felt that we were living in an era of incipient revolt and that we ourselves were mortally involved in this revolt against a social system and code of conduct and morality which, for convenience sake, may be referred to as bourgeois Victorianism. We did not initiate this revolt. When we went up to Cambridge, its protagonists were Swinburne, Bernard Shaw, Samuel Butler in *The Way of All Flesh*, and to some extent Hardy and Wells. We were passionately on the side of these champions of freedom of speech and freedom of thought, of common-sense and reason. We felt that, with them as our leaders, we were struggling against a religious and moral code of cant and hypocrisy which produced and condoned such social crimes and judicial murders as the condemnation of Dreyfus. People of a younger generation who from birth have enjoyed the results of this struggle for social and intellectual emancipation cannot realize the stuffy intellectual and moral suffocation which a young man felt weighing down upon him in Church and State, in the "rules and conventions" of the last days of Victorian civilization. Nor can those who have been born into the world of great wars, of communism and national socialism and fascism, of Hitler and Mussolini and Stalin, of the wholesale judicial murders of their own fellow-countrymen or massacres of peasants by Russian communists, and the slaughter of millions of Jews in gas-chambers by German nazis, these younger generations can have no notion of what the long-drawn out tragedy of the Dreyfus case meant to us. Over the body and fate of one obscure, Jewish captain in the French army a kind of cosmic conflict went on year after year between the establishment of Church, Army, and State on the one side and the small band of intellectuals who fought for truth, reason, and justice, on the other. Eventually the whole of Europe, almost the whole world, seemed to be watching breathlessly, ranged upon one side or other in the conflict. And no one who was not one of the watchers can understand the extraordinary sense of relief and release when at last the innocence of Dreyfus was vindicated and justice was done. I still think that we were right and that the Dreyfus case might, with a slight shift in the current of events, have been a turning point in European history and civilization. All that can really be said against us was that our hopes were disappointed.

It is true that in a sense "we had no respect for traditional wisdom" and that, as Ludwig Wittgenstein complained, "we lacked reverence for everything and everyone." If "to revere" means, as the dictionary says, "to regard as sacred or exalted, to hold in religious respect", then we did not revere, we had no reverence for anything or anyone, and, so far as I am concerned, I think we were completely right; I remain of the same opinion still—I think it to be, not merely my right, but my duty to question the truth of everything and the authority of everyone, to regard nothing as sacred and to hold nothing in religious respect. That attitude was encouraged by the climate of scepticism and revolt into which we were born and by Moore's ingenuous passion for truth. The dictionary, however, gives an alternative meaning for the word "revere"; it may mean "to regard with deep respect and warm approbation." It is not true that we lacked reverence for everything and everyone in that sense of the word. After questioning the truth and utility of everything and after refusing to accept or swallow anything or anyone on the mere "authority" of anyone, in fact after exercising our own judgment, there were many things and persons regarded by us with "deep respect and warm approbation": truth, beauty, works of art, some customs, friendship, love, many living men and women and many of the dead.

The young are not only ruthless; they are often perfectionist; if they are intelligent, they are inclined to react against the beliefs, which have hardened into the fossilized dogmas of the previous generation. To the middle-aged, who have forgotten their youth, the young naturally seem to be not only wrong, but wrong-headed (and indeed they naturally often are); to the middle-aged and the old, if they are also repectable, the young seem to be, not only wrong, but intellectually ill-mannered (and indeed they often are). In 1903 we were often absurd, wrong, wrong-headed, ill-mannered; but in 1903 we were right in refusing to regard as sacred and exalted, to hold in religious respect, the extraordinary accomplishment of our predecessors in the ordering of life or the elaborate framework which they had devised to protect this order. We were right to question the truth and authority of all this, of respectability and the establishment, and to give our deep respect and warm approbation only to what in the establishment (and outside it) stood the test and ordeal of such questioning.

It will be remembered that Maynard's *Memoir*, in which he analyses the state of our minds (and Moore's) when we were undergraduates, starts with an account of a breakfast party in Bertrand Russell's rooms in Cambridge, at which only Russell, Maynard, and D. H. Lawrence were present. Lawrence was "morose from the outset and said very little, apart from indefinite expressions of irritable dissent, all the morning." And in a letter to David Garnett, Lawrence referred to this visit of his to Cambridge as follows:

My dear David,

Never bring Birrell to see me any more. There is something nasty about him like black beetles. He is horrible and unclean. I feel I should go mad when I think of your set, Duncan Grant and Keynes and Birrell. It makes me dream of beetles. In Cambridge I had a similar dream. I had felt it slightly before in the Stracheys. But it came full upon me in Keynes and Duncan Grant. And yesterday I knew it again in Birrell . . . you must leave these friends, these beetles, Birrell and Duncan Grant are done for ever. Keynes I am not sure . . . when I saw Keynes that morning in Cambridge it was one of the crises of my life. It sent me mad with misery and hostility and rage . . ."

Maynard, starting from the breakfast party and this letter, examines the question whether there was in fact something in Lawrence's judgment, some justification for his horror and rage against "us", whether we were horrible and unclean and black beetles. His account of us, his dissection of our spiritual and intellectual anatomy, leads him to conclude that there was something in what Lawrence said and felt—there was a "thinness and superficiality, as well as the falsity, of our view of man's heart," we were "water-spiders, gracefully skimming, as light and reasonable as air, the surface of the stream without any contact with the eddies and currents underneath." In this, and indeed in the whole of the *Memoir*, Maynard confuses, I think, two periods of his and of our lives. When our Cambridge days were over, there grew up in London during the years 1907 to 1914 a society or group of people which became publicly known as Bloomsbury.

Later in my autobiography I shall have to say a good deal about Bloomsbury, the private nature and the public picture. Here all I need say is that Bloomsbury grew directly out of Cambridge; it consisted of a number of intimate friends who had been at Trinity and King's and were now working in London, most of them living in Bloomsbury.

Lawrence's breakfast party took place in 1914 or 1915. The people to whom he refers are not the undergraduates of 1903, but Bloomsbury, and a great deal of what Maynard wrote in his *Memoir* is true of Bloomsbury in 1914, but not true of the undergraduates of 1903. In 1903 we had all the inexperience, virginity, seriousness, intellectual puritanism of youth. In 1914 we had all, in various ways or places, been knocking about the world for ten or eleven years. A good deal of the bloom of ignorance and other things had been brushed off us. *Principia Ethica* had passed into our unconscious and was now merely a part of our super-ego; we no longer argued about it as a guide to practical life. Some of us were "men of the world" or even Don Juans, and all round us there was taking place the revolt (which we ourselves in our small way helped to start) against the Victorian morality and code of conduct. In 1914 little or no attention was paid to Moore's fifth chapter on "Ethics in relation to Conduct", and pleasure, once rejected by us theoretically, had come to be accepted as a very considerable good in itself. But this was not the case in 1903.

Moore and the Society were the focus of my existence during my last years at Cambridge. They dominated me intellectually and also emotionally, and did the same to Lytton Strachey and to Saxon Sydney-Turner. We were already intimate friends, seeing one another every day, before we were elected and got to know Moore well, but Moore's influence and the Society's gave, I think, increased depth and meaning to our relationship. I daresay to a good many people with whom we came into superficial contact we seemed, not without reason, unpleasant. Trinity was such a large college that, when we were up, one soon formed one's own circle of friends and acquaintances and rarely troubled or was troubled by those who belonged to other sets. The hostility of the ordinary man to the scholar and intellectual was therefore a good deal less important and less noticeable in Trinity than in some of the small colleges. It was only on nights of bump-suppers and similar drunks and celebrations that an intellectual or anyone who looked like an intellectual had to be careful to keep out of the way of the "bloods" and athletes. But I think that we—and Lytton in particular—got a special measure of dislike and misprision from the athletes and their followers—"the little men in waistcoats", as Thoby Stephen called them—on the rare occasions when our paths happened to converge. There was some reason, as I said above, for their finding us unpleasant, for we were not merely obviously much too clever to be healthy "good fellows", we were arrogant, supercilious, cynical, sarcastic, and Lytton always looked very queer and had a squeaky voice.

It must be admitted that it was not only among the toughs and bloods that we were unpopular. I still possess some letters, written in 1901 and full of the uncompromizing ferocity of youth which show this. During my second year at Trinity I had become friendly with a fellow scholar, himself an intellectual of considerable powers. When after a time he seemed to avoid me, I asked him the reason, and he explained his position in letters from which the following are extracts:

> "I cannot endure the people I meet in your rooms. Either they or I had to go, and as I was the newest and alone I waived my claim to the older friends and the majority. Strachey . . . &c. are to me in their several ways the most offensive people I have ever met, and if I had continued to meet them daily, I could not be answerable for anything I might do . . . I am not what is known as religious, but I was not going to associate with people who scoffed and jeered at my religion: fair criticism given in a gentlemanly way I do not mind. But the tone of Strachey and even you on matters of religion was not gentlemanly to me . . . I have never been in your rooms without someone coming in whom I do not like, usually Strachey . . . I always spoke to you as a friend to a friend, except when Strachey was with you. Silence is then safer."

As one grows older or even more as one grows old it is easy and pleasant to make either of two mistakes in one's memories of those few years when in one's early twenties one lived in a Cambridge college. One can foolishly idealize and sentimentalize youth and the young, and so oneself as young. Or one can do the opposite and join the many angry old men who are enviously exasperated by the young and therefore remember only the stupidities and humiliations of their own youth. There is a certain amount of truth in each of these views or visions so that either accepted absolutely and unmodified by the other is simply false. My own experience is that I have never again been quite so happy or quite so miserable as I was in the five years at Cambridge from 1899 to 1904. One lived in a state of continual excitement and strong and deep feeling. We were intellectuals, intellectuals with three genuine and, I think, profound passions: a passion for friendship, a passion for literature and music (it is significant that the plastic arts came a good deal later), a passion for what we called the truth.

What made everything so exciting was that everything was new, anything might happen, and all life was before us. We looked before, not after, and our laughter, continual and sincere, was not fraught with pain, just as our pain was as pure as our laughter. We lived in extremes—of happiness and unhappiness, of admiration and contempt, of love and hate. I might any day or hour or minute turn a corner and find myself face to face with someone whom I had never met before but who would instantly become my friend for life. I might casually open a book and find that I was reading for the first time *War and Peace*, *The Brothers Karamazov*, *Madame Bovary*, *Hedda Gabler*, *Urn Burial*, or *The Garden of Proserpine*. I might wake up tomorrow morning and find that I could at last write the great poem that fluttered helplessly at the back of my mind or the great novel rumbling hopelessly in some strange depths inside me. We all wanted to be writers ourselves, and what added to our excitement was that we could share our ambitions, our beliefs, our hopes, and fears. By "we" I mean pre-eminently Lytton Strachey, Saxon Sydney-Turner, Thoby Stephen, and myself.

The hates, contempts, miseries were as violent—almost as exciting—as the loves, friendships, admirations, ecstacies. We were arrogant, wrongheaded, awkward, oscillating between callowness and sophistication. We were convinced that everyone over twenty-five, with perhaps one or two remarkable exceptions, was "hopeless", having lost the élan of youth, the capacity to feel, and the ability to distinguish truth from falsehood. We were not angry young men in any sense; that psychology of a much later age was alien to ours. Intellectually we were terribly insolent, being contemptuous of, not angry with, authority and stupidity, so that no doubt to those whom we did not like or who did not like us we must have been insufferable.

Here I must again recall the fact, briefly mentioned already, that this period of our early manhood, perhaps the most impressionable years of one's life, was an age of revolution. We found ourselves living in the springtime of a conscious revolt against the social, political, religious, moral, intellectual, and artistic institutions, beliefs, and standards of our fathers and grandfathers. We felt ourselves to be the second generation in this exciting movement of men and ideas. The battle, which was against what for short one may call Victorianism, had not yet been won, and what was so exciting was our feeling that we ourselves were part of the revolution, that victory or defeat depended to some small extent upon what we did, said, or wrote. After the 1914 war, and still more after Hitler's war, the young who are not conservatives, fascists, or communists are almost necessarily defeatist; they have grown up under the shadow of defeat in the past and the menace of defeat in the future. It is natural, inevitable that they should suffer from the sterility of being angry young men. Our state of mind was the exact opposite.

There was no shadow of past defeat; the omens were all favourable. We were not, as we are today, fighting with our backs to the wall against a resurgence of barbarism and barbarians. We were not part of a negative movement of destruction against the past. We were out to construct something new; we were in the van of the builders of a new society which should be free, rational, civilized, pursuing truth and beauty. It was all tremendously exhilarating.

And no doubt, looking back after fifty years and two world wars and the atomic age and Mussolini, Hitler, Stalin, and Russian communism, no doubt terribly naïve. Of course, we were naïve. But age and hindsight unfairly exaggerate and distort the naïvety of youth. Living in 1900 and seeing the present with no knowledge of the future, we had some grounds for excitement and exhilaration. The long drawn out, crucial test of society and politics in the Dreyfus case had not yet ended in decisive defeat for the old régime, but the "pardoning" of Dreyfus foreshadowed their final defeat and the reinstatement of Dreyfus six years later. And what made the Dreyfus case so terribly exciting, so profoundly significant, was that this judicial murder of an obscure Jewish army officer, this trial and conviction of an unimportant captain in the 21st regiment of artillery in France, became, as I said above, a struggle of European and later almost of cosmic importance. One felt that gradually everyone in the world had become involved in it, that everyone was becoming consciously implicated in the struggle between right and wrong, justice and injustice, civilization and barbarism. The court-martial in the Cherche-Midi prison, the degradation ceremony on the parade-ground with the soldiers drawn up in a great square and Dreyfus raising his arms and crying out to them: "Soldiers, I am innocent! It is an innocent man who is being dishonoured. Vive la France! Vive l'Armée!", the crowd hissing and shouting: "A mort! A mort! Kill him! Kill him!", the imprisonment on the Ile du Diable—these events which were contemporary events of our own lifetime assumed symbolic import like the trial and death of Socrates and the scene in the prison at Athens—"Crito, we owe a cock to Asclepius"—or even that other trial before Pontius Pilate in Jerusalem and all the crowd hissing and shouting: "Let him be crucified! His blood be on us, and on our children."

I do not think we were wrong in feeling this tremendous significance of the Dreyfus case. It was, what we felt it to be, a struggle between two standards of social and therefore of human value. Two world wars and millions of Dreyfuses murdered by Russian communists and German nazis do not prove us to have been wrong; they merely show that any hope in 1904 that the world might become permanently civilized has not been fulfilled. One should perhaps recall the men and books which in those days of our Cambridge youth filled us with admiration, enthusiasm, hope, for they show the deep currents of revolt operating in the society of our time, and they reveal not only autobiographically our personal psychology, but also—what is more important—the historical psychology of an era.

There was in Trinity an old-established Shakespeare Society which met, I think, weekly to read aloud the plays of Shakespeare. Not content with this, we founded a new society, the X Society, for the purpose of reading plays other than Shakespeare. We read the Elizabethans and the Restoration dramatists with immense pleasure, but we also read two contemporaries, Ibsen and Shaw. The plays of these two writers gave us something over and above the aesthetic pleasure which we got from the poetry of *The Duchess of Malfi*, *Volpone*, or *The Maid's Tragedy*, or the intellectual pleasure which the wit of *The Way of the World* gave us. The poetry, the work of art in *The Wild Duck* or *Hedda Gabler* or *The Master-Builder* gave us and still give me profound pleasure, a pleasure which can rightly be distinguished as purely aesthetic. The dramatic genius, the humour, and the verbal wit in *Arms and the Man* or *You Never Can Tell* gave us and to some extent can still give me great intellectual pleasure. But in all these plays, and pre-eminently in Ibsen's, there was something else, something extraordinarily exciting which belonged to the immediate moment in which we lived and yet went down into the depths of our beliefs and desires and the great currents of history. Not only in such plays as *The Doll's House* or *Ghosts*, but in the strange symbolic words and action of *The Master-Builder* or *The Wild Duck* or *Rosmersholm*, the cobwebs and veils, the pretences and hypocrisies which suppressed the truth, buttressed cruelty, injustice, and stupidity, and suffocated society in the nineteenth century, were broken through, exposed, swept away. When Brack said: "Good God!—people don't do such things.", when Hilda says: "But he mounted right to the top. And I heard harps in the air. My—my Master-Builder!", when Relling says: "Bosh!" to Molvik's: "The child is not dead but sleepeth", we felt that Ibsen was revealing something new in people's heads and hearts—in our heads and hearts—and that he was giving us hope of something new and true in human relations and that he was saying "Bosh!" to that vast system of cant and hypocrisy which made lies a vested interest, the vested interest of the "establishment", of the monarchy, aristocracy, upper classes, suburban bourgeoisie, the Church, the Army, the stock exchange.

We did not think Bernard Shaw to be nearly as great a dramatist as Ibsen. He lacked the poetry, if that is the right word, the creative imagination, which makes *The Wild Duck*, *Hedda Gabler*, *Rosmersholm*, *John Gabriel Bjorkman* so moving, quite apart from any "message" that they had for our generation. But Shaw did have a message of tremendous importance to us—and to the world—in the years 1899 to 1904. It was the same kind of message which I have tried to describe above as coming to us from Ibsen's Brack, Hilda, and Relling. I still possess a copy of Vol. II of *Plays: Pleasant and Unpleasant* published by Grant Richards in 1901; I have written in it my name and the date, showing that I bought it in November 1902. In the preface Shaw answers his liberal critics who had attacked him for striking "wanton blows" at idealism, religion, morality, the "cause of liberty" in *Arms and the Man* and others of his plays. I will quote his own words, because they state contemporaneously, in November 1902, far better than I can today that message which made us recognize him with enthusiasm as one of our leaders in the revolutionary movement of our youth.

> "... idealism, which is only a flattering name for romance in politics and morals, is as obnoxious to me as romance in ethics or religion. In spite of a Liberal Revolution or two, I can no longer be satisfied with

fictitious morals and fictitious good conduct, shedding fictitious glory on robbery, starvation, disease, crime, drink, war, cruelty, cupidity, and all the other commonplaces of civilization which drive men to the theatre to make foolish pretences that such things are progress, science, morals, religion, patriotism, imperial supremacy, national greatness and all the other names the newspapers call them. On the other hand, I see plenty of good in the world working itself out as fast as the idealists will allow it; and if they would only let it alone and learn to respect reality, which would include the beneficial exercise of respecting themselves, and incidentally respecting me, we should all get along much better and faster. At all events, I do not see moral chaos and anarchy as the alternative to romantic convention; and I am not going to pretend I do merely to please the people who are convinced that the world is only held together by the force of unanimous, strenuous, eloquent, trumpet-tongued lying."

The novels which a man reads throw light upon his psychology and the psychology of his generation. Very few of the illustrious dead among English novelists meant much to us. In 1903 I read Fielding and Richardson rather because I thought they should have been read, Jane Austen and the Brontës because they gave me pleasure, aesthetic and intellectual. Thackeray and Dickens meant nothing to us or rather they stood for an era, a way of life, a system of morals against which we were in revolt. We were unfair to them and misjudged them aesthetically, as I recognized when much later in life I came to read them again. It was the curious satire of *Crotchet*

Castle, so suave and yet so sharp, which struck a note in the past which we could appreciate as in harmony with our mood. Of contemporaries the first to mean something to us was George Meredith. I don't think I ever *liked* him as much as the others of my generation did, for there seemed to be something unreal and phoney in his artificiality. But he appealed to us as breaking away from the cosmic and social assumptions of Thackeray and Dickens, as challenging their standards of morality. I am not sure now that he did and today I feel that we were almost certainly also wrong about Henry James. I have explained already the immense influence which he had upon us. Up to a point we were right and the influence was justified, for the niceties and subtleties of his art and his psychology belonged to the movement of revolt. But he was never really upon our side in that revolt.

There were two novelists, amazingly different from each other, who were very definitely upon our side and whom we recognized with enthusiasm, not merely as writers and artists, but as our leaders. *The Way of All Flesh* by Samuel Butler was published in 1903, a year after his death. We read it when it came out and felt at once its significance for us. The other was Thomas Hardy. *The Return of the Native* and *The Mayor of Casterbridge* seemed to us great novels, and, though we probably overvalued them, we were not far wrong. But those books and still more *Tess of the D'Urbervilles* and *Jude the Obscure* had another importance besides the artistic. The outcry against them came from those who supported what we thought the most degraded and hypocritical elements in Victorianism. *Tess of the D'Urbervilles* was published in 1891 and in the preface to the new edition of 1895 Hardy himself dealt with these critics. He pointed out that "a novel is an impression, not an argument", and he was not arguing a case in *Tess*. Some of those who objected to his novel were "genteel persons . . . not able to endure something or other" in the book. Some were "austere" persons who considered that certain subjects are not fit for art; some objected to a woman who had an illegitimate child being made the heroine of a respectable novel and still more to her being rather provocatively labelled by Hardy in his subtitle a "pure woman". Looking back over half a century to this, in these spacious days when solemn judges give solemn judgments that *Ulysses* and *Lady Chatterley's Lover* are not indecent and can be read innocuously by babes and sucklings as well as by pornographic elderly gentlemen, it is almost impossible to believe that in 1900 *Tess* was widely condemned as an immoral book. *Tess* and Hardy were themselves a cause of the change, just as were Ibsen and Shaw, and that was why we, who felt ourselves to be so much involved in this struggle of ideas and ideals, regarded all three of them as in a sense our leaders.

There is another name which I must add to these three, and I am afraid that I will appear ridiculous to practically everyone of later generations by admitting it. For it is Algernon Charles Swinburne. Late at night in the May term, I like to remember, Lytton, Saxon, Thoby Stephen, Clive Bell, and I would sometimes walk through the Cloisters of Nevilles Court in Trinity and looking out through the bars at the end on to the willows and water of the Backs, ghostly in the moonlight, listen to the soaring song of innumerable nightingales. And sometimes as we walked back through the majestic Cloisters we chanted poetry. More often than not it would be Swinburne:

"From too much love of living,
From hope and fear set free,
We thank with brief thanksgiving
Whatever gods may be
That no life lives for ever;
That dead men rise up never;
That even the weariest river
Winds somewhere safe to sea."

"We shift and bedeck and bedrape us,
Thou art noble and nude and antique."

"Thou hast conquered, O pale Galilean; the world has grown grey from thy breath;
We have drunk of things Lethean, and fed on the fulness of death."

It all sounds, no doubt, silly and sentimental and, what so many people think even more deplorable, so terribly out of date. Of course we were silly and sentimental at the age of twenty, and I do not think there is anything admirable in this kind of crudity and naïvety of the young. Yet the unfledged foolishness is partly due—and was in our case due—to the enthusiasm, the passion with which one sees and hears and thinks and feels when everything in the world and in other people and in oneself is fresh and new to one. This passion is, I think, admirable and desirable, and that is one reason why it irritates us when age and experience, bringing disappointment and boredom, have blurred our sensations, dimmed our beliefs, and castrated our desires.

As to the out-of-dateness of Swinburne and *Dolores* and *The Garden of Proserpine*, and so of the young men and the nightingales, I am not much concerned or troubled by it. I have always thought that to make a fuss or a song about being up-to-date is the sign of a weak mind or of intellectual cold feet or cold heart. Every out-of-date writer of any importance was once modern, and the most modern of writers will some day, and pretty rapidly, become out-of-date. For us in 1902 Tennyson was out-of-date and we therefore underestimated his poetry; today another fifty years has evaporated much of his datedness, and his stature as a poet becomes more visible. I daresay that we overestimated Swinburne's poetry,[1] but I have no doubt that it is generally underestimated today. If I wanted to chant a poem to nightingales singing at midnight and in moonlight, I might still choose the *Garden of Proserpine* or some other poem of Swinburne's and I am sure that it would stand up to the ordeal as well as any song the Sirens have sung. That is no mean or common achievement and would show that there is some poetry in the poem.

Swinburne was something of a legend and symbol to us in the early nineteen hundreds. Immured for the last thirty years of his life by Theodore Watts-Dunton in the grim bourgeois "residence", The Pines, at the bottom of Putney Hill, "his life was 'sheltered' like that of a child," as his biographer, Sir Edmund Gosse, wrote, "and he was able to concentrate his faculties upon literature and his

[1] But not very much, for we retained a certain sense of criticism. We used, as a kind of parlour game, to draw up a Tripos List of all the world's writers of all ages. I still possess one of these Lists compiled by us in 1902. Swinburne is placed in Class I, Division 3, below Browning who is placed in Class 1, Division 2. But we never chanted Browning at midnight to the nightingales.

dreams without a shadow of disturbance." His poetry is a kind of distilled lyricism and in this bears some resemblance to Greek lyric poetry, to Pindar and the choruses in Sophocles and Aristophanes, in which sense and sound become one and well up in song. And physically his tiny body, "light with the lightness of thistledown", seemed to be the perfect ethereal envelope for the lyricist. Living in Putney from 1894 and during my Cambridge years, I had a gleam of reflected glory from the poet. Nearly every morning he walked up Putney Hill and over Wimbledon Common and occasionally I saw him doing this. And once when I was having my hair cut in a shop near Putney station, the door opened and everyone, including the man cutting my hair, turned and looked at the tiny, fragile-looking figure in a cloak and large hat standing in the doorway. I remember very vividly the fluttering of the hands and fear and misery in the eyes. No one said anything, and after a moment the little figure went out and shut the door. "That", said the barber, "is Mr. Swinburne." I have one other memory. Our doctor at Putney was a Dr. White, who had played rugger at half-back for England. He was also Watts-Dunton's, and therefore Swinburne's, doctor. He told me that he was sometimes summoned by Watts-Dunton to see the poet. The interview took place in the dining-room in Watts-Dunton's presence. Swinburne sat at a long table and could rarely be induced to say anything, but all the time his little hands played an inaudible tune on the dining-table as though upon a piano. Watts-Dunton, a rather sinister figure, one often saw in Putney. When I took my dog for a walk, I would sometimes meet him in Putney Park Lane, then a very rural lane, arm-in-arm with a beautiful young lady, with whom we had a distant acquaintance, the enormous-eyed, almost Pre-Raphaelite looking, Clare Reich.

To return to Trinity, I have never been what is called "a good mixer", but I have always felt great interest in and often liking for all kinds of different persons. At Trinity, I had two quite distinct circles of intimate friends. One, which I have so far been dealing with and describing, consisted of intellectuals and scholars with Lytton Strachey, Saxon Sydney-Turner, Thoby Stephen, Maynard Keynes, and G. E. Moore at the centre and at varying distances from the centre Clive Bell, J. T. Sheppard, R. G. Hawtrey and A. R. Ainsworth. Two other persons moved erratically in and out of this solar system of intellectual friendship, like comets, Morgan Forster (E. M. Forster) and Desmond MacCarthy. They were both older than I, Morgan by two years and Desmond by three. Later in life when I returned from Ceylon in 1911 and lived in London, I became much more intimate with them, and I shall have more to say about them when I get to that period of my life. Morgan, I suppose, was still up at King's when I first knew him, though not in my last two years. We did not see very much of him, but he was a fascinating character and what I knew of him I liked immensely. He was strange, elusive, evasive. You could be talking to him easily and intimately one moment, and suddenly he would seem to withdraw into himself; though he still was physically there, you had faded out of his mental vision, and so with a pang you found that he had faded out of yours. He was already beginning to write his early Pan-ridden short stories and *A Room with a View*. You always felt in him and his conversation the subtlety and sensibility together with the streak of queer humour which you always also feel in his books. Lytton nicknamed him the Taupe, partly because of his faint physical resemblance to a mole, but principally because he seemed intellectually and emotionally to travel unseen underground and every now and again pop up unexpectedly with some subtle observation or delicate quip which somehow or other he had found in the depths of the earth or of his own soul. His strange character and our early relationship are shown, I think, in the following letter which he wrote to me just before I left England for Ceylon in 1904:

Harnham, Monument Green,
Weybridge.

14/11/04
Dear Woolf

I nearly was at Cambridge yesterday, but it didn't come off. I don't think, though, that I really wanted to see you again.

This letter is only to wish you godspeed in our language, and to say that if you ever want anything or anything done in England will you let me know. It's worth making this vague offer, because I'm likely always to have more time on my hands than anybody else.

I shall write at the end of the year. I know you much less than I like you, which makes your going the worse for me.

Yours ever
E. M. Forster.

In 1905 Morgan sent me an inscribed copy of *Where Angels Fear to Tread*; he has crossed out the title and written above it "Monterians".

Desmond MacCarthy had already gone down when I came up to Trinity in 1899, but in my last years I got to know him quite well as he was a great friend of Moore's and used fairly often to come up and stay with him. Desmond in youth was, I think, perhaps the most charming man that I have ever known. His charm was so much a part of his living person, the tone of his voice, the turn of his sentence, the tolerant or affectionate smile, the wrinkled forehead, the sagacious eye with the humorous gleam in it which reminded me, even when he was young, of the eye of a knowing old dog who understands and appreciates all his master's jokes and can make just as good jokes himself, that sixty years later it is, alas, hopeless to try to convey even a shadow of it to anyone who did not know him. The first time I met him was one weekend when he came up to stay with Moore. He had just returned from a kind of old-fashioned Grand Tour of Europe and he gave us immensely long descriptions of his journeys, particularly in Greece and Turkey. Such pyrotechnic displays or set pieces are usually the most boring type of conversation, but one could listen enchanted by Desmond as a raconteur until one was tired out, not by him, but by pleasure and laughter. He was one of those rare intellectuals who can talk to anyone in any place or "walk of life" and to whom everyone can talk easily and affectionately. If you went into a tobacconist's to buy a packet of cigarettes with him, it would almost certainly be ten minutes before you came out accompanied to the door by everyone in the shop laughing and talking with him up to the last possible minute. He seemed to me, and to many others of his friends, in those days to have the world at his feet. He had wit, humour, intelligence, imagination, a remarkable gift of words, an extraordinary power of describing a character, an incident, or a scene. Surely, one thought, here is in the making a writer, a novelist of the highest quality. As a human being he remained the same to the end, but as a writer he never achieved anything at all of what he promised. This is not the place to explore the reasons for his failure, interesting though they are, for they belong to a later period of his life and mine, and will therefore be discussed more appropriately when I come to the account of my and of his middle age. In my Cambridge days I remember him only as someone upon whom the good fairies appeared to have lavished ever possible gift both of body and of mind. For he was very strong and athletic. He used to play fives with Moore and me and, though I was not a bad player, he was very much better, playing a tremendously fast game and hitting with great power.

So much for the circle of my friends who were scholars, dons, "intellectuals". But I moved in a second circle which was almost the exact opposite of this. It was essentially heterogeneous. The kernel of it, so far as I was concerned, consisted of three men, each almost completely different from the other: Harry Gray, who had been at St. Paul's, Alan Rokeby Law, who had been at Wellington, and Leopold Colin Henry Douglas Campbell, who had been at Eton. My friendship with them, which outlasted Cambridge, began through Gray. At St. Paul's I knew him only by sight for he was on the science side. Somehow or other our paths crossed at Trinity and I went to see him, and it was in his rooms that I met Law and Campbell with whom he had become intimate. Gray was tall, very thin, and graceful, with extraordinarily neat, long hands; his head was very small and all his features small; his face was absolutely without any colour in it except a faint tinge of yellow. Though he was intelligent and very likable, his character in most ways was as colourless as his face. What attracted me in him was that, although lively and affectionate and interested, he seemed devoid of anything approaching passion. He was absorbed in two things, but with an almost impersonal absorption, medicine and music. He was taking the Science Tripos and in later life became a first-class surgeon. He was already, as an executant, a first-class pianist. His playing was brilliant, but singularly impersonal and emotionless, and, when he was not working, he would usually be found playing the piano. It was characteristic of him that he was usually playing Chopin and I used to listen with considerable pleasure and even excitement to the cold and limpid fountains of rhythm and melody which he made Chopin and the piano produce. I liked Gray and he liked me, but in a curiously impersonal way. Years afterwards, when I came back from Ceylon and he had become a distinguished surgeon, he asked me whether I would like to come and see him perform an operation. I have always thought that one should never refuse an experience, so I went to see him take a large growth out of an elderly man's inside. The operation astonished me: it lasted for a very long time and for much of it Gray used only his hands in the man's inside, exerting considerable force. Watching his hands, I was continually reminded by their movements of the way in which he used to play Chopin.

Law was a tiny little man with the palest of straw-coloured hair. At the age of twenty he had the face of a rather puzzled old man and I think that he must have looked much the same at the age of two or even at birth. He and his clothes were wonderfully dapper and he was so tidy that nothing in his rooms might be moved an eighth of an inch from the place appointed for it by him within the room and the universe. He was very affectionate and loyal, so that among his friends all his geese were swans. As he was both conventional and respectable, he would naturally have thought me a blasphemous and predatory goose; he turned me into a swan by seeing that, though brainy and queer, I was a good fellow. I went and stayed with him in his home in Ripley and was fascinated by a glimpse into a stratum of society into which I had never penetrated. He lived an only child with his parents in a small country house and garden typical of the Surrey of those days. His parents were typical too; conservative, conventional, commonplace, they belonged to a not unimportant part of the middle middle-class backbone of nineteenth-century England. Their intense respectability was strongly tinged with snobbery towards both those above and those below them in the social scale. They had not too much and not too little of everything, including wealth and brains. I cannot remember what Mr. Law was or had been, but the family climate was that of the Church and Army, not usually soaring above the rank of an Archdeacon or Colonel, though Mrs. Law had a cousin who was a General and had survived the Boer War without the discredit earned by most of the British commanders in South Africa. They must have produced Rokeby rather late in their life; he was not merely the centre of their universe, he was their universe. Their universe was in no sense mine; our standards of value were in most things antagonistic; I had no sympathy for the stuffiness of their postulates, beliefs, ideals. Yet beneath the carapace which class, religion, and public school had formed over their brains and souls, there was something in them and in Rokeby that I liked and found interesting, perhaps because it was so different from anything under my carapace. I made them rather uneasy, but they accepted me as a friend of Rokeby. He died quite young, a few years after he went down from Trinity, and with his death their universe collapsed. They sent me a copy of a biographical memoir of him which they had privately printed.

Leopold Colin Henry Douglas Campbell, who later in life became Leopold Colin Henry Douglas Campbell-Douglas and still later Lord Blythswood, belonged again to a class completely different from that of Law. He was a Scottish aristocrat, tracing his descent from Sir Colin Campbell of Lochow in the fourteenth century, Colin Campbell of Blythswood in the seventeenth century, and the Douglases of Douglas-Support in the eighteenth century. His ancestors were a long line of soldiers; his uncle, the first Lord Blythswood, his father, and one of his brothers were all colonels of the Scots Guards. The aristocratic class and way of life to which Leopold Campbell belonged could scarcely have been more different from or more antagonistic to mine, but I have always enjoyed plunging, with a shudder and shiver, into a strange and alien society of people, as into an icy sea. To aristocratic societies I know that I am ambivalent, disliking and despising them and at the same time envying them their insolent urbanity which has never been more perfectly described than in *Madame Bovary*. Here is Flaubert's description of the old and young men whom Emma met when she was invited to stay with the Marquis d'Andervilliers—so like the old and young men whom I met at lunch in the Campbells' house in Manchester Square in 1903:

> Leurs habits, mieux faits, semblaient d'un drap plus souple, et leurs cheveux, ramenés en boucles vers les tempes, lustrés par des pommades plus fines. Ils avaient le teint de la richesse, ce teint blanc que rehaussent la pâleur des porcelaines, les moires de satin, le vernis des beaux meubles, et qu'entretient dans sa santé un régime discret de nourritures exquises . . . Ceux qui commencaient à vieillir avaient l'air jeune, tandis que quelque chose de mûr s'étendait sur le visage des jeunes. Dans leurs regards indifférents flottait la quiétude de passions journellement assouvies; et, à travers leurs manières douces, perçait cette brutalité particulière que communique la domination de choses à demi faciles, dans lesquelles la force s'exerce et oû la vanité s'amuse, le maniement des chevaux de race et la société des femmes perdues.

Leopold's father was a General who had just got through the Boer war without either credit or discredit; he looked like and was essentially a General who had been Colonel of the Scots Guards. Leopold was in many ways true to type, but in some ways a mutation or sport. He talked of huntin', shootin', and ridin', and even in later life frequented "la société des femmes perdues", but at Eton he suddenly became virulently infected with religion and determined to become a High Church parson. There must have been a gene in the family producing embryos with a tendency to become clergymen rather than colonels, because Leopold's uncle, the second Lord Blythswood, went up to Trinity College, Cambridge, and took orders. My experience is that almost everyone, if you really get to know them, is a "curious character", and Leopold was no exception. I liked him and he liked me, although—or perhaps to some extent because—we had so very little in common. He had many of the infuriating prejudices and affectations of his class and caste, but he was affectionate and, what is rare, a man of real good-will. He had a singularly open mind in some directions and an appreciation of people; though my mother was in almost every way different from the ladies whom he knew, when he got to know my family, he became very fond of her and she of him. What attracted me in him was the spontaneous gaiety and benevolence in his nature and an unusual mental curiosity. At Cambridge he was preparing to take Holy Orders and in due course he became a High Church parson. Why exactly he went through this exacting process and lived the whole of his life as curate, vicar, or rector of parishes in towns or villages, I never really understood. His religion puzzled me though he never hesitated to talk quite freely to me about it. At a certain psychological level it was perfectly genuine, but the level was only just below the surface of his mind or soul. He took a kind of aesthetic pleasure in the paraphernalia of High Church services, the incense and genuflexions and all the rest of it, or perhaps it is truer to say that the Church was to him exactly what the Scots Guards were to his father, his uncle, and his brother. At any deeper level, religion and Christianity seemed to mean nothing, to have no relevance for him. I feel sure that, if he and I had been walking down Piccadilly and had suddenly come face to face with Jesus Christ, I should have recognized him instantly, but Leopold, if he noticed him at all, would have dismissed him as merely another queer-looking person.

My friendship with Leopold Campbell caused one curious incident. There was an undergraduate at Trinity of our year, whom I will call X and whom we both knew and rather liked. He was a slightly dim person and I was astonished and outraged to hear from more than one acquaintance that he was going about saying that I "sponged on Campbell". I was not yet, at the age of twenty, steeled to expect and ignore the malignancy of men or the spurns that patient merit from the unworthy takes, and I went round to the rooms of X on the first floor of a house on King's Parade terribly hurt and terribly angry. There followed a scene of violent emotion in which I exacted a grudging apology and left the room in such a rage that I fell down the stairs from the first to the ground floor.

I had to keep the Gray-Law-Campbell circle as far apart as possible from the circle of my intellectual friends as they were mutually suspicious and antagonistic. By the time I left Cambridge I had become very intimate with Thoby Stephen and Lytton Strachey and knew their families, and so the foundations of what became known as Bloomsbury were laid. Thoby's family seemed to a young man like me formidable and even alarming. When his father, Sir Leslie Stephen, came up to stay a weekend with him, Lytton and I were had in to meet him. He was one of those bearded and beautiful Victorian old gentlemen of exquisite gentility and physical and mental distinction on whose face the sorrows of all the world had traced the indelible lines of suffering nobility. He was immensely distinguished as a historian of ideas, literary critic, biographer, and the first editor of the *Dictionary of National Biography*. In each of these departments the distinction was not undeserved; his *History of English Thought in the Eighteenth Century* seventy-six years after it was published, according to Mr. Noel Annan, Provost of King's College, Cambridge, and a very professional, a very modern, and a very exacting critic, in his book *Leslie Stephen, His Thought and Character in Relation to his Time*, "still stands as a major contribution to scholarship and in a sense will never be superseded in its scope." Maynard Keynes, another expert and severe critic, also thought highly of this book. Stephen's literary criticism in *Hours in a Library* and his biographies, deeply etched by his personal prejudices and the conventional ethics of respectable Victorians, are of remarkable quality and are still readable and read. When I found myself, a nervous undergraduate, sitting opposite to this very tall and distinguished old gentleman in Thoby's rooms in Trinity Great Court and expected to make conversation with him—not helped in any way by Thoby—it seemed to me, as I said, formidable and alarming. What added enormously to the alarm was that he was stone deaf and that one had to sit quite near to him and shout everything one said to him down an ear-trumpet. It is remarkable and humiliating to discover how imbecile a not very imaginative, or even an imaginative, remark can sound when one shouts it down an ear-trumpet into the ear of a bearded old gentleman, six foot three inches tall, sitting very upright in a chair and looking as if every word you said only added to his already unendurable sorrows. However it must be said that this awkwardness and terror were gradually dissipated by him. He had immense charm and he obviously liked to meet the young and Thoby's friends. Unlike Henry James, he could see through our awkwardness and even grubbiness to our intelligence, and was pleased by our respect and appreciation. In the end we were all talking and laughing naturally (so far as this is possible down an ear-trumpet) and enjoying one another's company. This must have been about three years before Leslie Stephen died.

The basis of Mr. Ramsey's character in *To the Lighthouse*, was, no doubt, taken by Virginia from her father's character; it is, I think, successfully sublimated by the novelist and is not the photograph of a real person stuck into a work of fiction; it is integrated into a work of art. But there are points about it which are both artistically and psychologically of some interest. Having known Leslie Stephen in the flesh and having heard an enormous deal about him from his children, I feel pretty sure that, subject to what I have said above about the artistic sublimation, Mr. Ramsey is a pretty good fictional portrait of Leslie Stephen—and yet there are traces of unfairness to Stephen in Ramsey. Leslie Stephen must have been in many ways an exasperating man within the family and he exasperated his daughters, particularly Vanessa. But I think that they exaggerated his exactingness and sentimentality and, in memory, were habitually rather unfair to him owing to a complicated variety of the Oedipus complex. It is interesting to observe a faint streak of this in the drawing and handling of Mr. Ramsey.

I also met Thoby's two sisters, Vanessa and Virginia Stephen, when they came up to see him. The young ladies—Vanessa was twenty-one or twenty-two, Virginia eighteen or nineteen—were just as formidable and alarming as their father, perhaps even more so. I first saw them one summer afternoon in Thoby's rooms; in white dresses and large hats, with parasols in their hands, their beauty literally took one's breath away, for suddenly seeing them one stopped astonished and everything including one's breathing for one second also stopped as it does when in a picture gallery you suddenly come face to face with a great Rembrandt or Velasquez or in Sicily rounding a bend in the road you see across the fields the lovely temple of Segesta. They were at that time, at least upon the surface, the most Victorian of Victorian young ladies, and today what that meant it is almost impossible to believe or even remember. Sitting with them in their brother's room was their cousin, Miss Katherine Stephen, Principal of Newnham, with whom they were staying. But Miss Stephen was in her cousin's room for a tea-party, not in her capacity of cousin, but in her capacity of chaperone, for in 1901 a respectable female sister was not allowed to see her male brother in his rooms in a male college except in the presence of a chaperone. I liked Miss Stephen very much, but it could not be denied that she was a distinguished, formidable, and rather alarming chaperone. All male Stephens—and many of the females—whom I have known have had one marked characteristic which I always think of as Stephenesque, and one can trace it in stories about or the writings of Stephens of a past generation whom one never knew, like the judge, Sir James Fitzjames Stephen, and the two James Stephens of still earlier generations. It consisted in a way of thinking and even more in a way of expressing their thoughts which one associates pre-eminently with Dr. Johnson. There was something monolithic about them and their opinions, and something marmoreal or lapidary about their way of expressing those opinions, reminding one of the Ten Commandments engraved upon the tables of stone, even when they were only telling you that in their opinion it would rain tomorrow. And what was even more characteristic and Stephenesque was that usually over this monolithic thought and these monolithic pronouncements there played—if one dare use the word of these rather elephantine activities—a peculiar monolithic humour.

The Principal of Newnham had a liberal measure of this Stephen method of thinking and talking and it was not calculated to put at his ease a nervous young man who met her for the first time. Thoby, a most monolithic Stephen—his affectionate nickname, The Goth, fitted his great stature and monumental mind—was himself a little shy and quite incompetent to deal with a slightly sticky social situation. In such a situation he was inclined to sit silent, smiling tolerantly and deprecatingly. Vanessa and Virginia were also very silent and to any superficial observer they might have seemed demure. Anyone who has ridden many different kinds of horses knows the horse who, when you go up to him for the first time, has superficially the most quiet and demure appearance, but, if after bitter experience you are accustomed to take something more than a superficial glance at a strange mount, you observe at the back of the eye of this quiet beast a look which warns you to be very, very careful. So too the observant observer would have noticed at the back of the two Miss Stephens' eyes a look which would have warned him to be cautious, a look which belied the demureness, a look of great intelligence, hypercritical, sarcastic, satirical.

In the Stephen family there was a vein of good looks, particularly noticeable in the males. There was something very fine in Leslie Stephen's face and his nephew Jim (the famous "J.K."), the son of James Fitzjames, must have been a handsome man. The mother of Vanessa and Virginia was born Julia Jackson; when Leslie Stephen married her, she was the widow of Herbert Duckworth. Julia Jackson's mother was one of the six Pattle sisters whose great beauty was legendary. The Pattle genes which caused this beauty must have been extremely potent, for there is no doubt that it was handed on to a considerable number of descendants. It had some very marked individual characteristics—for instance, the shape and modelling of the neck, face, and forehead, and the mouth and eyes—which were and are traceable in the third and fourth generation. It was essentially female for, though it has certainly produced many lovely women, I cannot see any trace of it in the male descendants of the Pattles whom I have known. Julia Jackson inherited a full measure of the Pattle beauty, as one can see in the famous photographs by her famous aunt, Mrs. Cameron, herself one of the six Pattle sisters. All this is not irrelevant to my story. No one could deny that the Pattle sisters and their female descendants in the next generation were extraordinarily beautiful, but it was a beauty which was or tended to become rather insipid. It was, I think, too feminine, and not sufficiently female, and there was about it something which was even slightly irritating. Vanessa and Virginia had inherited this beauty, but it had been modified, strengthened, and, I think, greatly improved by the more masculine Stephen good looks. When I first met them, they were young women of astonishing beauty, but there was in them nothing of the saintly dying duck loveliness which was characteristic of some of their feminine ancestors. They were eminently Stephen as well as Pattle. It was almost impossible for a man not to fall in love with them, and I think that I did at once. It must, however, be admitted that at that time they seemed to be so formidably aloof and reserved that it was rather like falling in love with Rembrandt's picture of his wife, Velasquez's picture of an Infanta, or the lovely temple of Segesta.

While I was at Cambridge, I got to know Lytton Strachey's family a good deal better than I did Thoby Stephen's. The Strachey and Stephen families both belonged to a social class or caste of a remarkable and peculiar kind which established itself as a powerful section of the ruling class in Britain in the nineteenth century. It was an intellectual aristocracy of the middle class, the nearest equivalent in other countries being the French eighteenth century noblesse de robe. The male members of the British aristocracy of intellect went automatically to the best public schools, to Oxford and Cambridge, and then into all the most powerful and respectable professions. They intermarried to a considerable extent, and family influence and the high level of their individual intelligence carried a surprising number of them to the top of their professions. You found them as civil servants sitting in the seat of permanent under-secretaries of government departments; they became generals, admirals, editors, judges, or they retired with a K.C.S.I. or K.C.M.G. after distinguished careers in the Indian or Colonial Civil Services. Others again got fellowships at Oxford or Cambridge and ended as head of an Oxford or Cambridge college or headmaster of one of the great public schools.

Stephens and Stracheys were eminent examples of this social development, and, when I got to know them well, I was both interested and amused by the great difference in outlook and postulates of their circle from many of those in the circle from which I came. The Strachey family, as I said, belonged to exactly the same class as the Stephens, but, owing I suppose to its individual genes, it was in many ways extraordinarily different. When I first knew Lytton, they lived in a very large house in Lancaster Gate. I used sometimes to call formally on his mother, Lady Strachey, on Sunday afternoons, according to the strange custom of those prehistoric times, and sometimes I was asked to remain to Sunday supper. Lytton's sister Pippa, who was the most energetic and charming of women and remains so today, ageless in the eighties, decided that we must all be taught by her to dance Highland dances and I went to one or two evening parties at which about twenty young people, including the Miss Stephens, practised this difficult art under the lively and exacting tuition of Pippa.

I stayed with Lytton three years running in the summer in large country houses which his parents rented, once in Surrey, once in Essex, and once near Bedford. In this way I got to know his father and mother and all his brothers and sisters. They stand out in my memory as much the most remarkable family I have ever known, an extinct social phenomenon which has passed away and will never be known again. Lytton's father was Lieutenant-General Sir Richard Strachey, who, like his two brothers, had had an extraordinarily brilliant career in India. He was a remarkable product of his caste in nineteenth-century Britain, a man of immense ability whether in action or intellectually, for he attained eminence as an army officer, an administrator, an engineer, and a scientist. When I knew him he must have been eighty-five years old, a little man with a very beautiful head, sitting all day long, summer and winter, in a great armchair in front of a blazing fire, reading a novel. He was always surrounded by a terrific din which, as I shall explain, was created by his sons and daughters, but he sat through it completely unmoved, occasionally smiling affectionately at it and them, when it obtruded itself unavoidably upon his notice, for instance, if in some deafening argument one side or the other appealed to him for a decision. He was usually a silent man, who listened with interest and amusement to the verbal hurricane around him; he was extraordinarily friendly and charming to an awkward youth such as I was, and he was fascinating when now and again he was induced to enter the discussion or recall something from his past.

Lytton's mother, like his father, came from a distinguished Anglo-Indian family, being the daughter of Sir John Grant. She was a large, tall, rather ungainly woman who often appeared to be completely detached from the world around her. She would walk into a room in a kind of dream-like way, gaze uncertainly about her and then walk out again. I used to think that she had come in to try to find something which she had forgotten and then, when she was in the room, forgot what it was she had forgotten. She would often sit at the head of the Strachey table apparently unconscious of her children's babel of argument, indignation, excitement, laughter. This absent-mindedness or distraction was, however, rather deceptive. She was, in fact, tremendously on the spot whenever she gave her attention to anything. She was passionately intellectual, with that curiosity of mind which the Greeks rightly thought so important. She had a passion for literature, argument, and billiards. I was very fond of her and got on well with her, and she liked me, I think. The houses which they took for the summer in the country were necessarily large and always had a billiard-table, and I played many games of billiards with her. She was a magnificent reader of poetry and in the evenings she would read aloud to Lytton and me for hours. The last time that Virginia and I saw her was when she was old and blind, sitting one summer evening under a tree in Gordon Square. We went and sat down by her, and somehow or other we got on the subject of Milton's *Lycidas*, which at St. Paul's we had to learn by heart. She recited the whole of it to us superbly without hesitating over a word. The beauty of the London evening in the London square, the beauty of the poem of that old blind Londoner who sat on summer evenings 300 years before in Lincoln's Inn Fields composing *Paradise Lost*, the beauty of Lady Strachey's voice remain one of the last gentle memories of a London and an era which vanished in the second great war.

In 1902 the Strachey family consisted of five sons and five daughters, female and male alternating down the family thus: Elinor (Mrs. Rendel); Dick, in the Indian Army; Dorothy (later Madame Simon Bussy); Ralph, married, in the East Indian Railway; Pippa, later secretary of the Fawcett Society; Oliver, married, in the East Indian Railway; Pernel, later Principal of Newnham; Lytton; Marjorie; James. Their ages ranged from Elinor's forty-two years to James's fifteen; their father,

Sir Richard, had been born in the reign of George III two years after Waterloo, and their grandfather, Edward Strachey, far, far back in 1774. These years and dates are not irrelevant to a description and understanding of Lytton and his family. I felt that whereas I was living in 1902, they were living in 1774–1902. At dinner someone might casually say something which implied that he remembered George IV (which he might) or even Voltaire or Warren Hastings, and certainly to Lytton the eighteenth century was more congenial and, in a sense, more real than the nineteenth or the twentieth. The atmosphere of the dining-room at Lancaster Gate was that of British history and of that comparatively small ruling middle class which for the last 100 years had been the principal makers of British history.

At supper on Sunday evenings in Lancaster Gate or still more in the country houses in the summer the number of Stracheys present was to a visitor at first bewildering. In London the family consisted of Sir Richard, Lady Strachey, Dorothy, Pippa, Lytton, Marjorie, James, and Duncan Grant, Lady Strachey's nephew. In the country there was always a large influx of the married sons and daughters, with wives, husbands, and children. The level of intelligence in each son and daughter and in the father and mother was incredibly, fantastically high. They were all, like their mother, passionately intellectual, most of them with very quick minds and lively imaginations. All of them, I suspect, except the two eldest, must have been born with pens in their hands and perhaps spectacles on their noses. Their chief recreation was conversation and they adored conversational speculation which usually led to argument. They were all argumentatively very excitable and they all had in varying degrees what came to be known as the Strachey voice. I have said something about it in describing Lytton. It had a tremendous range from deep tones to high pitched falsetto. When six or seven Stracheys became involved in an argument over the dinner table, as almost always happened, the roar and rumble, the shrill shrieks, the bursts of laughter, the sound and fury of excitement were deafening and to an unprepared stranger paralysing. And these verbal typhoons were not confined to literary discussions and the dinner table. I was once playing croquet with Lytton, Marjorie, and James when I stayed with them in a house near Bedford and a dispute broke out between the three Stracheys over some point in the game. I stood aside waiting until the storm should subside. The noise was terrific. The back of the house, which was, I think, early Georgian, looked down with its, say, eighteen windows upon the lawn where we were playing. By chance I looked up at the house and was delighted to see a Strachey face at each of the eighteen windows watching the three furious gesticulating figures and listening, I think appreciatively, to the noise and excitement.

In my family we were very energetic and, in a mild way, adventurous. I have always liked to do something new or experimental, and my brother Herbert and I thought nothing of setting off on our bicycles at short notice to ride from Putney to Edinburgh or of deciding on a bicycle tour in the Shetlands. Lytton was by our standards very unadventurous, but when I stayed with him in the house near Bedford it struck me that it would be amusing to hire a canoe at Bedford and canoe down the Ouse to Ely where the Cam flows into it and then canoe up the Cam to Cambridge. I induced Lytton, with some difficulty, to agree to do this with me. I reckoned to take about four or five days for the whole expedition. We set off in brilliant, hot, sunny August weather. The Ouse was amazingly beautiful. There was some legal dispute over navigation which had gone on for years, and for years all the locks on the river were closed to boats or barges. The river was entirely deserted except for our canoe and innumerable birds. It seemed to be very high, the water almost topping the banks. One paddled or floated down an immense, lovely, interminable tunnel of blue water, feathery reeds and meadowsweet, bright green fields and bright blue sky, accompanied by an unending escort of flashing kingfishers. In one respect it was a pretty strenuous expedition, since, the locks being closed, every now and again, when one came to a lock, one had to lift the canoe out of the water and carry it round to float it again below the lock. In the evening we tied up and found an inn to stay the night in and it was all extremely pleasant. But the second day it started to rain, heavily and pitilessly, and at Huntingdon, I think, we abandoned the canoe and took a car to Peterborough, where we stayed the night. Next morning it was still raining and dejectedly we returned to Bedford by train.

I stayed up for five years at Trinity. I had come up with a vague intention of eventually becoming a barrister, having as a small child announced that I would be what Papa was and drive every morning in a brougham to King's Bench Walk. But it was not long before I changed my mind and decided to go into the Civil Service, expecting as a scholar of Trinity to be high enough up in the examination to get a place in the Home Civil Service.

Meanwhile as a classical scholar I had to take the Classical Tripos. In my day you got your B.A. degree on the Classical Tripos, Part I, and to take Part II was a luxury usually indulged in only by those who thought they might subsequently get a Fellowship. In my third year I got a First Class in the Tripos, but in the third division, which disappointed the authorities, so the Master, the great Dr. Montagu Butler, wrote to me, since they thought I should have been placed in the first division. I decided that I would spend my fourth year reading for Part II of the Classical Tripos, Greek philosophy. I did even worse in that than in Part I, for I only got a Second Class. I spent my fifth year reading for the Civil Service examination. This was a great mistake. For the Civil Service you needed to be crammed in as many subjects as possible besides your main subject (in my case classics). In twelve months you crammed into your head as much as possible of subjects called, for instance, political economy, political science, logic. The university was ill-equipped for cramming and one really ought to have gone down and delivered oneself into the hands of a London crammer who knew how to treat the human brain like the goose who is to become pâté de foie gras.

Compared with most scholars I did little work at Cambridge, if work means going to lectures, reading, and stuffing your head with what will give you a high place in an examination. I hate lectures, and, as at Trinity the authorities did not insist upon scholars attending them punctiliously, I went to few. I read voraciously both in Greek and Latin and in English and French, but it was not the kind of diet which wins you very high marks in an examination. I am quite good at exams, but the truth is that I was a really first-class classical scholar when I came up from St. Paul's to Trinity, but nothing like as good when I took Part I of the Classical Tripos. When I took the Civil Service examination, I could read Greek and Latin fluently, as I still can, but I had forgotten all the paraphernalia of syntax and writing Greek and Latin compositions. The result was that I got poor marks in the classical papers in which I should have amassed most of my marks and so did extremely badly. The best that I could hope for was a place in the Post Office or Inland Revenue. I was over age for India. I felt that I could not face a lifetime to be spent in Somerset House or in the Post Office, so I decided to take an appointment in the Colonial Service, then called Eastern Cadetships. I applied for Ceylon, which was the senior Crown Colony, and I was high enough up on the list to get what I asked for. I found myself to my astonishment and, it must be admitted, dismay in the Ceylon Civil Service.

Looking back I can see that the dismay was natural, but unnecessary. I am glad that I did not go into the Home Civil and did go into the Ceylon Civil Service. My seven years in Ceylon were good for me, and, though they gave me a good deal of pain, they gave me also a good deal of pleasure—and a great deal of pleasure as a memory of things past. But I am glad too that I lived the kind of life at Trinity which was mainly the reason why I did not do well in the examinations. It was, I think, a civilized life both intellectually and emotionally. My intellect was kept at full stretch, which is very good for the young, by books and the way I read them and by friends and their incessant and uncompromising conversation. The emotion came from friendship and friends, but also from the place, the material and spiritual place, Trinity and Cambridge. I must try, before I end this chapter and this period of my life, to say something of the effect of the institution, the college and the university with the surrounding aura of the town and the country, an effect partly of place and partly of history and spirit.

The attitude of a person to the institutions, collectivities, groups, herds, or packs with which he is or has been associated throws considerable light upon his character and upon the hidden parts of it. Biographers and autobiographers, as a rule, say little about it and many people are reticent about their "loyalties" other than that which Johnson significantly described as the last refuge of a scoundrel. (I propose to use the word loyalty to cover a person's emotions or reactions of a positive and appreciative nature to any group or institution.) When I try to look objectively into my own mind, I detect feelings of loyalty to: my family; "race" (Jews); my country, England in particular, and the British Empire generally; places with which I have been connected, such as Kensington and London (born and bred), counties, Middlesex and Sussex, where I have lived, Ceylon, Greece; school; Trinity and Cambridge. Some of the evidence for the existence of these loyalties is curious, but unmistakable. For instance, in the case of St. Paul's, Cambridge University, England, and the British Empire, if they win in any game or sport, I get distinct pleasure, and pain if they lose. I also want Sussex and Middlesex to win in county cricket, but here for some more complicated reasons I also want Yorkshire and Gloucestershire to win, and there is a conflict of loyalties.

The quality of these loyalties differs profoundly in the

different cases. In the case of places like London, England, Sussex, Ceylon I love them for their material beauties or excellences, but also for spiritual qualities, memories, traditions, history. Still more significant is the fact that in some cases, particularly family, race, and school, my feelings are ambivalent. The first wounds to one's heart, soul, and mind are caused in and by the family, and deep down unconsciously one never forgets or forgives them. One loves and hates one's family just as—one knows and they know—one is loved and hated by them. Most people are both proud and ashamed of their families, and nearly all Jews are both proud and ashamed of being Jews. There is therefore always a bitterness and ambivalence in these loyalties. My feelings towards St. Paul's school are ambivalent, but in a different way. I hated its physical ugliness, its philistinism, the slow, low torture of boredom that crept over one as one sat hour after hour in the stuffy class-room listening to the bored voice of the bored master. I still hate it, and yet I have at the same time affection for and pride in it. I get irrational pleasure from the knowledge that I was taught in a school which has now existed for 450 years and that it was connected with Erasmus. I like to think that I went to the same school as Milton, Marlborough, and Pepys. I do not feel the same towards another Old Pauline, Field-Marshal Lord Montgomery, as I would if I had been to Eton or Croydon Secondary Modern School; I wish I thought him as great a general as he thinks himself and, quite irrationally, I feel personal regret that he says such silly things.

My loyalty to Trinity and Cambridge is different from all my other loyalties. It is more intimate, profound, unalloyed. It is compounded of the spiritual, intellectual and physical inextricably mixed—the beauty of the colleges and Backs; the atmosphere of long years of history and great traditions and famous names; a profoundly civilized life; friendship and the Society. Soon after I went to Ceylon, Desmond MacCarthy wrote me a long letter in which among other things he described how he had been up to Trinity to the great annual feast, Commem. I will quote what he says, because I think it gives an extraordinarily keen taste of Trinity College and why, even in its most absurd moments, it meant something to Desmond and to all of us. The letter is dated March 23, 1905:

> "I have only been up to Cambridge twice since I saw you. Once for Commen: I was McTaggart's guest, and I stayed with him. I didn't enjoy the dinner very much —I overate and overdrank to meet his sense of the occasion. We didn't have very good talk. I relish his wit; but he doesn't enjoy my jokes, at least I feel he doesn't, so I cant make them. The scene was just what you remember, neither very easy nor dignified—and yet it made a claim on one's feelings, as standing for something. "The college" "Trinity"—do these words mean much to you? They do mean something to me. Yet there are several things which mean so much more that I find it hard sometimes to believe that I have *any* esprit de corps—(corps d'esprit an old horse dealer, with an excellent French accent, called it as I travelled up in the train). Then the Master got up, holding his glass of wine in both hands and swaying solemnly from side to side in a way that in itself was a benediction, proposed the guests in a speech of admirable blandness and effortlessness and nothingness and as I listened I

felt the glamour of success. How fine it is that the college should send out men who become ministers and judges and bishops and how very gratifying it would be to come down as the bigwig of the evening and make a most splendid speech in reply—all this sublimated by the rosy mist of port. And I looked down the table and caught sight of the Davieses[1] their faces set in deep disgust.

"Then McT and Theodore and I went to the reception at the Lodge. The Master stood at the top of the stairs and welcomed us—received us, wrapped us round with romantic ceremonial hospitality. He was the Master of Trinity—the leaders of their generations were there—I was a brilliant young man—it was an Occasion.

"Then we went to Jackson's[2] at home. You know the scene—clouds of tobacco smoke, a roar of conversation—dozens of whist tables with lighted candles on each—clay pipes—boxes of cigars, a piano and someone singing God knows what—the tune coming in gusts as through an opening and shutting door above the babel of voices. Then songs with choruses—school songs, the various representatives of different schools gathering round the piano in turn and shouting with defiant patriotism—then music hall songs—imagine Parry on a sofa in the window, singing and enjoying the fun of the thing—"Oh my darling, oh my darling,

[1] Theodore Llewelyn Davies in the Treasury, and his brother Crompton, a solicitor. They were both very good-looking and charming, brilliant intellectually, high-principled, austere.

[2] Henry Jackson, O.M., Regius Professor of Greek and one of the senior Fellows of Trinity, a great scholar and a great character.

oh my darling Clementine"—Then came the event of the evening for me. I got on the sofa with Strachey and had a good talk. First about the people in the room—beginning with the sour-faced Montague,[1] standing a little way off, whom I wish always to take gently above the elbow and inform that he is not another Dizzy. We talked for a long time about the Society, angels, brothers, and embryos—and felt as tho' we were geting things clear—at least I did—and we agreed."

To stay up at Cambridge for a fifth year, as Saxon and I did, is a curious experience, a little melancholy, though the gentle melancholy was not unpleasant. It was a kind of twilight existence, a respite, a waiting for the business of life to begin. Practically everyone of one's own year had gone down, and though I saw more and more of Moore and Maynard Keynes, it was often a solitary life. I came up for six weeks in the "long vacation" and that was still more solitary, for only a very few people did that. In some ways, one felt Cambridge, the essence of Cambridge, more intimately in the deserted courts than in term time. I came up ostensibly to read for my exam and I suppose that I did do a certain amount of work. It was a hot summer and the long summer days passed slowly away. There was a strange man called Barwell whom I knew and who also was up "for the long". He was said to be a descendant of that Richard Barwell, Member of Council in India, who supported Warren Hastings against Philip Francis. Philip Francis said of Richard Barwell that "he is rapacious without industry, and ambitious

[1] Edwin Montagu, later Secretary of State for India in Lloyd George's 1916 Coalition Government.

without an exertion of his faculties or steady application to affairs. He will do whatever can be done by bribery and intrigue; he has no other resource." But Warren Hastings said of him that "his manners are easy and pleasant." His descendant, the Barwell whom I knew, seemed to have inherited the qualities which Hastings saw in his ancestor, and I never detected any sign of the sinister characteristics which Francis describes. He was a good deal older than I and had gone down from Trinity before I came up. But he was quite often about the place, and his manners were indeed easy and pleasant. He was what is called a man of the world, almost the first of that curious human species whom I got to know fairly well. I could never be and never shall be, I know well, a man of the world, and I rather despise them with at the same time a sneaking envy of them—but I often get on quite well with them. I got on quite well with Barwell. He liked good food, good wine, and good conversation. He used to take me, an Irish baronet whose name I no longer remember, and a B.A. called Maclaren, in a punt up the river and there we tied up under the willows and ate chicken, drank Burgundy, and talked through the long summer evenings like scholars and men of the world. And as darkness fell we punted slowly and silently back to Trinity.

Slowly the days and weeks passed away, and the Civil Service examination was on us in one of the hottest of London summers. The torture was prolonged for the exam went on for about three weeks. I remember coming out into the blazing sunshine in Burlington Street with Saxon quite often to find Thoby Stephen waiting for us so that we could lunch together, and then back again into the examination room in Burlington House to answer or not answer absurd questions about Logic, or Political Science, or Political Economy.

But even a Civil Service exam ends at last, and on August 29th, 1904, I set out with Maynard Keynes on a walking tour from Denbigh. We walked with knapsacks on our backs from Denbigh through Bettws-y-Coed, Beddgelert, and Pwllheli to Aberdaron. There we stayed for a few days with Charlie[1] and Dora Sanger. Then we walked to and climbed Snowdon, ending our tour finally at Carnarvon. Maynard was an extraordinarily good companion and even in those days had a passion for gambling, which I shared. We talked all day and in the evening did what we could to satisfy our gambling passion by playing bezique. Maynard kept the score and I remember him working out the result in the train on our way back to London, the result being that I won a few shillings off him.

The result of the exam was a considerable shock, but somehow or other I had learnt very early in life not to worry about things, to make up my mind quickly, and not to waste one's energies and emotions in regrets. I saw at once that I was not going to sit in Somerset House for the rest of my life and that therefore I must go to Ceylon.

[1] Charles Percy Sanger was a remarkable man. He was an Apostle, but a good deal older than I. He did brilliantly at Trinity and then became a barrister and was at the top of his profession as a conveyancer. He was a gnomelike man with the brightest eyes I have ever seen and the character of a saint, but he was a very amusing, ribald, completely sceptical saint with a first-class mind and an extremely witty tongue, a mixture which I never came across in any other human being. He wrote a very interesting pamphlet about Emily Brontë, which I published in the Hogarth Press, *The Structure of Wuthering Heights*.

The next few weeks, until I finally walked up the gangway of the P. and O. *Syria* at Tilbury, passed away quickly in the atmosphere of a kaleidoscopic dream. I bought tropical suits and a topi at the Army and Navy Stores; I took riding lessons in the Knightsbridge Barracks, a terrifying procedure, and in Richmond Park, which was a pleasant antidote. I passed my medical examination at any rate triumphantly for the doctor complimented me on having the cleanest feet of anyone he had examined that morning—"though," he added, "I am bound to say that that is not saying very much." I travelled up to Stonehaven near Aberdeen and spent a night with my brother Harold on a farm where he was learning agriculture. I spent a night with Moore in Edinburgh. I went through ceremonies of farewell with Desmond, who gave me the Oxford Press miniature edition of Shakespeare and Milton in four volumes which have accompanied me everywhere ever since, with Morgan Forster, Lytton, and Saxon. I had a farewell dinner with Thoby Stephen and his sisters Vanessa and Virginia in Gordon Square. I felt just as I did when as a small boy at school in Brighton I stood in Brill's Baths and looked down at the water so far below and nerved myself for the high dive. I got ready everything which I was to take with me to Ceylon, which included ninety large volumes of the beautiful eighteenth century edition of Voltaire printed in the Baskerville type and a wire-haired fox-terrier. At last I dived; the waters closed over me; I took the train to Tilbury Docks.

INDEX

Books by Leonard Woolf
available in paperback editions
from Harcourt Brace Jovanovich, Inc.

SOWING: AN AUTOBIOGRAPHY OF THE YEARS 1880 TO 1904

GROWING: AN AUTOBIOGRAPHY OF THE YEARS 1904 TO 1911

BEGINNING AGAIN: AN AUTOBIOGRAPHY OF THE YEARS 1911 TO 1918

DOWNHILL ALL THE WAY: AN AUTOBIOGRAPHY OF THE YEARS 1919 TO 1939

THE JOURNEY NOT THE ARRIVAL MATTERS: AN AUTOBIOGRAPHY OF THE YEARS 1939 TO 1969